HEREDITY

An Introduction to GENETICS

ABOUT THE AUTHOR

A. M. Winchester is Professor of Biology at Colorado State College in Greeley, Colorado and a former consultant on the Biological Science Curriculum Study which was carried on at the University of Colorado under the sponsorship of the American Institute of Biological Sciences. After completing his undergraduate training at Baylor University, he received his M.A. and Ph.D. degrees at the University of Texas, where he studied genetics under the Nobel prize winner, H. J. Muller. He has done post-graduate work at the University of Chicago, Harvard University, the University of Michigan, and the University of Munich. His teaching experience has included assignments at Baylor University, Stetson University, Lamar Technological College, University of Vermont, University of South Carolina, University of Virginia, Emory University, and Appalachian State College.

In addition to numerous research articles, Dr. Winchester is the author of three widely used textbooks in the fields of genetics, biology, and zoology. He is also the author of a popularized book on human heredity.

Dr. Winchester was formerly president of the Florida Academy of Science and the Academy Conference of the American Association for the Advancement of Science. He is a fellow of the Association and is listed in *Who's Who in America*. He is a member of the Genetics Society of America, the American Society of Human Genetics, the Eugenics Society of America, Sigma Xi, and Beta Beta Beta.

COLLEGE OUTLINE SERIES

HEREDITY

An Introduction to GENETICS

Second Edition

A. M. WINCHESTER

BARNES & NOBLE, INC.   NEW YORK
PUBLISHERS   •   BOOKSELLERS   •   SINCE 1873

©
SECOND EDITION, 1966

Copyright, 1961, 1966
BY BARNES & NOBLE, INC.

Seventh Printing, 1967

L. C. Catalogue Card Number: 66–25040

All rights reserved

This is an original College Outline (Number 58) in the original College Outline Series. It was written by a distinguished educator, carefully edited, and manufactured in the United States of America in keeping with the highest standards of publishing.

PREFACE

This book is intended as an outline of the fundamental principles of genetics, presented in a form which can be easily comprehended by both the student and the general reader. As a part of the College Outline Series, one of its aims is to provide the student of genetics with a supplement to his classroom lecture notes and his textbook. In addition, this outline should be of value to those in various fields who require some knowledge of the fundamentals of heredity. For the medical student, who must often undertake a study of medical applications of genetics without previous training in the subject, it should serve as a valuable reference. To those interested in the behavioral sciences, where the impact of heredity upon human activities and relationships is of great importance, this book will serve as a useful guide. Practicing physicians, psychologists, and counselors who must deal with problems of marriage and child-bearing will find this volume offers them the factual information they require. Members of the legal profession and other persons involved with problems of law will find that questions of paternity and even criminal investigations may often be solved through genetic means. Parents and prospective parents will find interesting reading here as they attempt to understand and estimate the influences of heredity on their children. Finally, any person who would further his education through self-teaching will find the subject of genetics one of the most interesting branches of science, and one which easily lends itself to such a learning process.

The photographs and drawings used to illustrate the text are all made by the author unless otherwise indicated by credit lines. For a fuller account of the subject, the interested reader may wish to refer to any one of the standard textbooks listed in the Bibliography. Particularly to be recommended, of course, is the author's own textbook, *Genetics*, published by Houghton Mifflin and Co. A word of appreciation is due to these publishers for their cooperation and for permission to use some of the illustrations from this textbook.

A. M. Winchester

TABLE OF CONTENTS

CHAPTER	PAGE
1. Genetics—Its Methods and Applications	1
The Applications of Genetic Knowledge	1
Methods of Genetic Study	4
2. Early Development of Genetic Knowledge	12
Evidences of Ancient Understanding of Genetics	12
The Greek Philosophers	13
The Dawn of Modern Concepts	14
Nineteenth-Century Controversies	17
Gregor Mendel and the Development of the Gene Concept	21
The Early Twentieth Century	22
3. The Physical Basis of Heredity	23
Genes and Chromosomes	23
Mitosis	25
Meiosis	31
Maturation of Animal Germ Cells	32
Fertilization	36
Meiosis and Gene Assortment	37
Meiosis and Interspecies Hybrids	38
Meiosis in Plants	40
4. The Monohybrid Genetic Cross	41
Mendel's Monohybrid Crosses	41
Mechanisms of the Monohybrid Cross	45
The Test Cross	48
Genes with Intermediate Expression	50
Genes Affecting Viability	52
5. The Dihybrid Genetic Cross	59
Mendel's Dihybrid Crosses	59
The Dihybrid Cross Diagram	60
More Complex Hybrid Crosses	61
Modified Dihybrid Ratios	62
Short Cuts in Obtaining Phenotypic Ratios	66

6. Probability — 70
- Applications to Genetics — 70
- Law of Coincident Happenings — 71
- The Binomial Method — 74
- Short Cuts to Expansion of the Binomial — 75
- Statistical Methods of Analysis of Results — 77

7. The Determination of Sex — 86
- Sexual Bipotentiality of Organisms — 86
- The Separation of the Sexes — 86
- Sex Determination by Chromosomes — 87
- Hormones and Sex Determination — 91
- Sex Intergrades — 94
- Variations in the Sex Ratio — 96
- Attempts at Predetermination of Sex — 98

8. Relation of Sex to Inheritance — 100
- Sex-linked Genes — 100
- Incompletely Sex-linked Genes — 107
- Holandric Genes — 107
- Sex-limited Genes — 108
- Sex-influenced Genes — 110

9. Multiple Alleles and Multiple Genes — 114
- Multiple Alleles — 114
- Multiple Genes (Polygenes) — 116
- Multiple Genes and Quantitative Characteristics — 118
- Use of the Standard Deviation — 121

10. Genetics of the Human Blood Groups — 126
- The ABO Blood Groups — 126
- Medico-legal Applications of Blood Group Inheritance — 131
- The Secretor Trait — 132
- The M and N Blood Antigens — 133
- The RH Blood Antigens — 133
- Other Blood Antigens — 138

11. Gene Linkage — 141
- Discovery of Linkage — 141
- Crossing Over Between Linked Genes — 142
- Genetic Applications of Crossover Data — 145
- Chromosome Mapping — 148
- Significance of Crossing Over in Selection — 152

12. Chromosome Aberrations — 156
- Aberrations Involving Portions of Chromosomes — 156
- Aberrations Involving Entire Chromosomes — 161

	Aberrations Involving Entire Sets of Chromosomes	162
	The Salivary Glands of Drosophila	163
	Value of Giant Chromosomes in Aberration Studies	166
13.	THE ROLE OF CHROMOSOMES IN SEX DETERMINATION	170
	Drosophila	170
	Hymenoptera (*Habrobracon*)	173
	Chromosomes and Sex in ZW Animals	174
	Plants with Separate Sexes	175
	Man	176
14.	GENE STRUCTURE	179
	The Chemical Nature of Genetic Material	179
	The Components of DNA	184
	The Gene Unit	187
	Cytoplasmic Inheritance	188
15.	GENE ACTION	193
	Human Hemoglobin	193
	Genes and Enzymes	198
	Enzymes in Neurospora	200
	An Enzyme Series in Man	203
	Transmission of Genetic Information	206
	Genetic Code	209
	Gene Mutations	210
	Control of Gene Activity	211
16.	GENE MUTATIONS	213
	The Nature of Mutations	213
	Methods of Mutation Detection	218
	Mutation Frequency	223
	The Artificial Induction of Mutations	227
	Cause of Natural Mutations	229
	Hardy-Weinberg Principle of Determining Frequencies of Mutant Genes	230
17.	RADIATION HAZARDS IN AN ATOMIC AGE	233
	Sources of High-Energy Radiation	233
	Measurement of Radiation	239
	Biological Effects of High-Energy Radiation	241
	Genetic Effects of High-Energy Radiation	247
	Radiation Tolerance Doses	251
	Exposures from Medical Use of X Rays	252
	The Thirty-Year Gonadal Dose	253
	The Atom Bomb Studies in Japan	253
	Reduction of Radiation Hazards	254

18. Heredity and Environment — 257
   Expressivity — 257
   Penetrance — 262
   Phenocopies — 263
   Twin Studies — 265

Index — 273

BIBLIOGRAPHY OF STANDARD TEXTBOOKS

The following list gives the author, title, publisher, and date of publication of the textbooks referred to in the table on pages xii–xiii.

Altenburg, Edgar. *Genetics*. Revised ed. New York: Henry Holt and Co., 1957.
Colin, Edward C. *Elements of Genetics*. 3rd ed. New York: McGraw-Hill, 1956.
Dodson, Edward O. *Genetics*. Philadelphia: W. B. Saunders Co., 1956.
Gardner, Eldon J. *Principles of Genetics*. 2nd ed. New York: John Wiley and Sons, 1964.
Herskowitz, Irwin H. *Genetics*. 2nd ed. Boston: Little, Brown and Co., 1965.
King, Robert C. *Genetics*. 2nd ed. New York: Oxford University Press, 1962.
Singleton, W. Ralph. *Elementary Genetics*. Princeton: D. Van Nostrand Co., 1962.
Sinnott, Edmund W., L. C. Dunn, and Theodosius Dobzhansky. *Principles of Genetics*. 5th ed. New York: McGraw-Hill, 1958.
Snyder, Laurence H., and Paul R. David. *The Principles of Heredity*. 5th ed. Boston: D. C. Heath and Co., 1957.
Srb, Adrian M., Ray D. Owen, and Robert S. Edgar. *General Genetics*. San Francisco: W. H. Freeman and Co., 1965.
Stern, Curt. *Principles of Human Genetics*. 2nd ed. San Francisco: W. H. Freeman and Co., 1960.
Whittinghill, Maurice. *Human Genetics and Its Foundations*. New York: Reinhold Publishing Corp., 1965.
Winchester, A. M. *Genetics*. 3rd ed. Boston: Houghton Mifflin Co., 1966.

QUICK REFERENCE TABLE

All numbers
See preceding page for

Chapter and Topic in This Outline	Altenburg	Colin	Dodson	Gardner	Herskowitz
1. Genetics—Its Methods and Applications	14–24	202–205 211–212	1–7 249–258	1–6	1–5
2. Early Development of Genetic Knowledge	361–363 441–448	1–10 52–64 208–211 214–215	273–283	6–14 24–30	$s1$–$s143$
3. The Physical Basis of Heredity	1–14 193–198	63–82	20–33	31–50	5–30
4. The Monohybrid Genetic Cross	25–39	11–36	8–19 82–87	15–18	31–41
5. The Dihybrid Genetic Cross	40–69	37–51 83–103	34–42 97–111	18–24	42–56
6. Probability	70-75 449–457	21–25	43–62	51–76	520–539
7. The Determination of Sex	97–101	139–145	63–65	100–103	102–115
8. Relation of Sex to Inheritance	101–105	165–180	65–81	103–130	90–101
9. Multiple Alleles and Multiple Genes	76–96 202–213	104–117	88–90 112–125	244–259 284–304	59–68
10. Genetics of the Human Blood Groups	214–231	108–117 343–351	90–96	204–205 249–253	185–198
11. Gene Linkage	160–202	118–138	126–143	77–99	116–148 355–362
12. Chromosome Aberrations	232–291	236–247	144–156 168–178	131–166	164–178 228–251
13. The Role of Chromosomes in Sex Determination	104–129	145–164	179–204	166–170	102–115
14. Gene Structure	393–425	216–220	205–223	250–283	252–305 330–338 369–382
15. Gene Action	336–390	220–225	258–272	196–229	383–472
16. Gene Mutations	14–15 292–307 328–347	225–236	157–167	171–195	149–163 189–227 306–316
17. Radiation Hazards in an Atomic Age	308–327	229–232	162–167	180–195	179–188
18. Heredity and Environment	426–440	148–156 181		262–273	473–491

xii

TO STANDARD TEXTBOOKS

refer to pages
complete titles of books

King	Singleton	Sinnott et al.	Snyder & David	Srb et al	Stern	Whitting-hill	Win-chester
53-73	1-15		3-10	505-509	1-6	1-15	1-20
	185-189	1-17	35 90-92 223-224				21-35
33-51	16-33	45-58	35-50 274-293	64-123	7-30 63-80	16-34	36-67
75-79	34-54	32-44 59-70 125-132	11-34	1-17	31-51 88-120 289-328	47-77	68-86 170-181
79-94	55-68	71-98	**51-79**	17-32		137-151	87-99
94-98	69-76	241-253 388-418	80-89	47-63 442-504	81-87 127-173	35-46 **94-107**	100-114
181-200	206-228	143-146 323-325	92-94	33-46	400-407 424-444	166-168	115-154
116-122	90-100	146-159	94-125 422-426	37-46	218-244	166-188 236-253	**155-169**
99-112	128-145 229-246	99-124	174-179 197-221	450-504	174-176 350-368	109-119 204-215	182-185 199-217
243-251	237-246	115-124	180-196	395-418	176-217 342-347	256-266	185-198
113-142	101-128	160-194	125-173 422-426	146-189	245-288	152-165	218-239
143-180	265-300	195-213	294-331	190-237 509-513	12-30	306-360	240-260 261-271
181-200	206-228	303-314	332-347	34-42	400-423	311-321	121-139
295-308	247-264 408-417	317-322 359-378	362-379 401-414	125-145 265-282		30-34	272-283
231-293	338-356	326-338	380-400	513-516	52-62	219-235	283-314
201-229 353-391	301-310	214-230	348-357 420-434	238-246 394-441	445-501	78-93 297-305 379-393	315-333
209-229	310-337	231-240	357-362	246-264	503-529	362-378	334-367
	76-89	7-31 133-142	415-418	516-540	530-607	264-294	368-426

HEREDITY

An Introduction to GENETICS

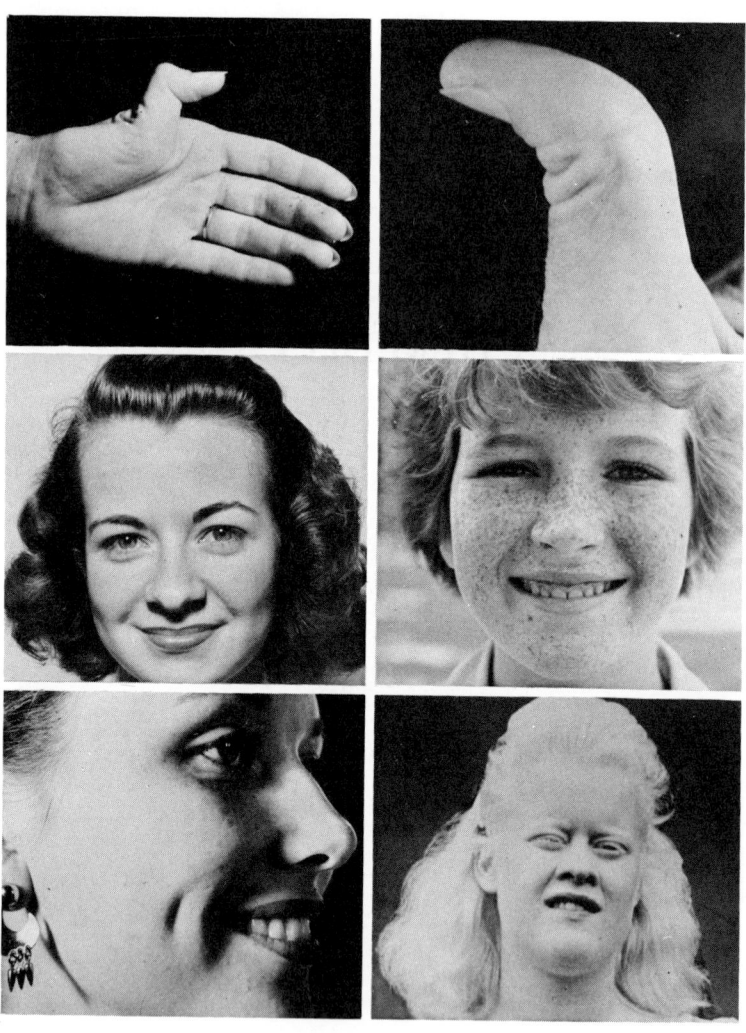

Some common inherited human characteristics. From left to right these are: hyperextensibility of the joints, hitch-hiker's thumb (ability to bend thumb back at an extreme angle), widow's peak (a point of hair extending downward in the center of the forehead), freckles, dimples in cheeks, albinism (reduced pigment in hair, skin, and eyes).

1: GENETICS—ITS METHODS AND APPLICATIONS

Genetics is the study of heredity—it is sometimes defined as the science that seeks to explain the similarities and differences that exist between organisms related by descent. Perhaps this definition is too trite and concise, but it serves to convey some concept of the primary subject matter involved. The word "gene" is used frequently in genetics as a designation for each of the small units of heredity within a cell, and it is through an expansion of this word that we derive the name of the science.

Genetics is a branch of the more inclusive science of biology, but geneticists also draw upon such related sciences as chemistry, mathematics, and physics in their efforts to ferret out all of the complex reactions which are involved in the transmission of inherited characteristics down through the generations. The science of genetics is attracting widespread interest because of its contributions to many phases of our lives. For example, we need to know more about genetics in order to evaluate the influence of high-energy radiations on future generations of man through alteration of genetic mechanisms.

THE APPLICATIONS OF GENETIC KNOWLEDGE

Genetics has proved to have numerous practical applications because man has learned to use the discoveries in many different fields. It is being used in such diverse areas as plant and animal breeding, medical diagnosis, and genetic counseling, and even in cases of law.

Plant and Animal Breeding. Genetics is an invaluable aid in man's efforts to improve domesticated animals and cultivated plants. The fat, beef-producing cattle on the western plains of the United States today are a far cry from the scrawny cattle that formerly grazed in that region. It has been largely through an in-

Fig. 1.1. Improvement of domestic animals through applications of genetic methods. The lean, long-horned steer on the left is a far cry from the fat bull at right that has been bred to produce beef. Likewise, the wild boar would have much less value today than its descendant, the Black Hampshire boar, in terms of meat production for market. (Courtesy U. S. Dept. of Agriculture.)

telligent application of the principles of genetics that this has been accomplished. Because of genetic breeding and selection the corn and wheat fields in the United States now yield over twice as much grain per acre as they did fifty years ago. One could not enjoy a delicious apple or a sweet, juicy Valencia orange had not the principles of genetics been applied to the production of these varieties of fruit. Without the agricultural improvements that genetics has brought about during the past century in the quality and quantity of yield of these crops, today's available land could not feed the present population.

Medical Applications. Heredity plays an important role in the development of many human afflictions, and a knowledge of heredity's exact role sometimes can be of great help in the prevention, diagnosis, and treatment of disease. Tuberculosis is caused by a germ infection, yet pedigree studies of many human families show that the predisposition to this disease is influenced

Fig. 1.2. Better oranges are produced through application of genetic methods. The sweet, juicy, thin-skinned, almost seedless orange at left is far superior, as far as man is concerned, to the sour, heavily seeded wild orange from which it was derived.

by heredity. Such information can be of value in attempts to prevent the disease. Individuals with this background can be warned of the environmental circumstances conducive to tubercular infection and can have more frequent checkups to detect possible early infection.

As an example of the use of genetics in the prevention of noninfectious diseases we can consider *xanthoma tuberosum*, which is characterized by the appearance of numerous nodules and tumors in the body and may involve the heart or blood vessels, sometimes with fatal results. The nodules develop because of an excess of cholesterol in the blood. If a person knows that a gene for this condition is present in his family, he can have frequent checks made on the cholesterol level of his blood. If the level becomes unduly high the afflicted individual should go on a low-cholesterol diet thus avoiding the dangers of the advanced stages of this abnormality.

Genetic Counseling. Closely related to the medical applications of genetics is the field of counseling. Most genetic counseling concerns human defects which lie in the province of medicine. There are a number of genetic counseling centers in the United States—for example, the Dight Institute for Human Genetics at the University of Minnesota and the Human Genetics Department at the University of Michigan. The typical clients of such centers are prospective parents who have in their family backgrounds some serious human abnormalities, such as clubfoot, mongolism, or harelip. These people would like to know the chances of such defects appearing in their children. Sometimes, when the risks seem to be too great, a couple may decide to forego having children of their own and rely upon adoption to have a family. In such a way many a human tragedy can be avoided. As the study of human genetics proceeds, more accurate information can be given because it is sometimes possible to detect carriers of harmful genes, even though the genes are not fully expressed. (See Chapter 4 for further details about this.)

Legal Applications. Many court cases today rely on geneticists for valuable testimony. Questions of disputed parentage may often be solved by identification of the blood types of the persons in question. Other inherited characteristics may also be of value. Divorce, custody of children, estate inheritance, and support of illegitimate children may involve problems which can be solved by genetic analysis.

METHODS OF GENETIC STUDY

The geneticist has various methods of study which he can use in investigations of problems of inheritance. These include experimental breeding, statistical analysis, cytology, and biochemical genetic study.

Experimental Breeding. This is the method through which most of our understanding of the principles of heredity has been derived. It consists of the crossing of organisms which differ with respect to certain inherited traits, followed by a careful tabulation of the offspring produced and an analysis of the results in an attempt to determine the method whereby these traits are transmitted. Certain forms of life are, obviously, more suited to this type of study than others. Human beings are not good subjects for they have the stubborn habit of wanting to choose their marriage partners for themselves and would not be amenable to suggestions that they have very large families in order to provide more reliable genetic ratios. Moreover, the human life cycle is so long and the maximum number of offspring so small that man would not be a good choice.

Four important factors must be considered when one chooses organisms for experimental breeding. These are:

1. *A short life cycle.* One could learn little about inheritance in elephants through experimental breeding because of their long life cycle. It would require hundreds of years to accumulate sufficient data to permit any accurate conclusions. Mice, however, are ready for breeding within six weeks after birth, and this is one factor which makes them a favorite for genetic breeding programs.

2. *A large number of offspring.* Genetics depends upon a statistical analysis of the results of crosses for its conclusions. Such studies require relatively large numbers in order to yield significant results. Hence, those forms of life are favored which give large numbers of offspring with each cross.

3. *Variation in inherited characteristics.* It would be impossible to learn anything about the inheritance of short hair in dogs if all dogs had short hair. Since there are many dogs, however, that have long hair we can cross these two variant forms and learn something about the method of inheritance of hair length. In all such studies, the forms of life usually chosen for experimental breeding are those that show considerable variation in the characteristics which are likely to be influenced by heredity.

4. *Convenience and economy.* This is an item of major importance, especially in animal breeding where the problem of feeding and caring for hundreds of animals can be very great if the animals are large and require much attention. It is customary to select some of the smaller animals for most such breeding experiments. Sufficient work has been done on larger forms to show that the pattern of inheritance is the same throughout the animal kingdom and the results obtained from the smaller types of animal can be applied to the larger types. The problem is somewhat simpler in the case of plants because there are many types of plant life which can be raised conveniently and cheaply. Much information about animal genetics has been derived through studies of plant breeding—in fact, the original discoveries of the method of inheritance of unit characteristics were made with plants.

The mouse happens to satisfy all of the above criteria. (For the experimental breeding of mammals the mouse has been used in genetic crosses more often than any other mammal.) There is

6 GENETICS—ITS METHODS AND APPLICATIONS

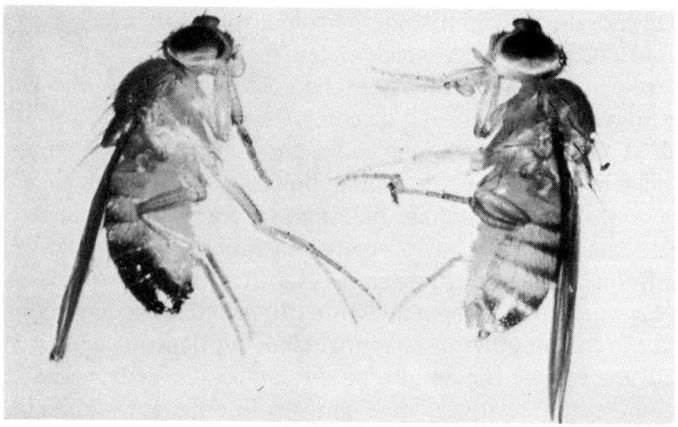

Fig. 1.3. The fruit fly, Drosophila melanogaster. *This little fly has been studied extensively by geneticists and much present-day knowledge of genetics has come from such studies. These are living flies as they appear when etherized for study and viewed under a microscope. The fly on the left is a male and the one on the right is a female.*

another animal, however, which satisfies the criteria even better and is used more extensively than the mouse. This is the little fruit fly, *Drosophila melanogaster*, which may be seen buzzing around fruit stands and garbage cans. This little gnat-sized insect has many features which are ideal from the standpoint of genetic breeding. Its life cycle is only ten to fifteen days and one female can produce more than a hundred offspring. Hundreds of inherited variations have been found. The eyes vary in color, size, and texture; the wings vary in size, curvature, and arrangement of the veins; the bristles of the body vary in arrangement, size, and length; the body varies in color and shape; there is hardly any part of the fly which does not show some inherited variations. Only a few cents worth of food and a small vial are necessary for raising hundreds of *Drosophila*, so that they are from every standpoint most economical.

The fruit fly would appear to have one great disadvantage—it is so very small—but with modern wide-field binocular microscopes it can be magnified so that every detail can be seen as clearly as would be possible if larger animals, such as cats or dogs, were so studied. Through the simple procedure of etherization the flies can be made to remain still while they are being examined and handled. They become quiescent within only a few seconds after they have been dropped into a bottle containing ether fumes and will remain quiet long enough to be handled and studied; they recover, moreover, within a few minutes without any signs of injury. All these factors have made *Drosophila* the most widely used animal for genetic breeding.

Statistical Analysis. After the offspring are obtained through experimental breeding and the results tabulated, we must then apply some of the principles of statistical analysis in order to secure a greater understanding of the genetic mechanisms which have been operating. So we must deal with ratios and the laws of probability in analyzing the results. Such methods can also be used in studies of inheritance in forms of life where experimental breeding is not practical. It is especially valuable in studies of human inheritance. While we cannot breed human beings with variant characteristics in order to learn how the characteristics are inherited, we can find cases where there have been marriages between persons having the characteristics in question and tabulate data about the offspring. Through a study of many such marriages we can assemble sufficient information concerning numbers of offspring to obtain reliable ratios.

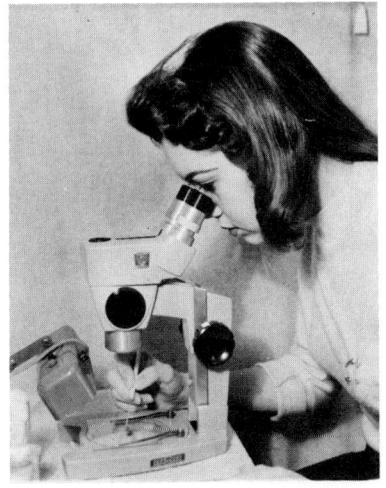

Fig. 1.4. *Fruit flies are small, but their most minute characteristics can be identified through the use of a modern stereo-microscope such as the one shown here.*

Family pedigrees may be constructed for such studies. These may be diagrammed as shown in Fig. 1.5. Males are represented by squares and women by circles. Marriage is indicated by horizontal lines connecting the marriage partners, and children are shown attached to a vertical line extending down from the horizontal line. The characteristic under investigation is indicated by

EAR LOBE PEDIGREE

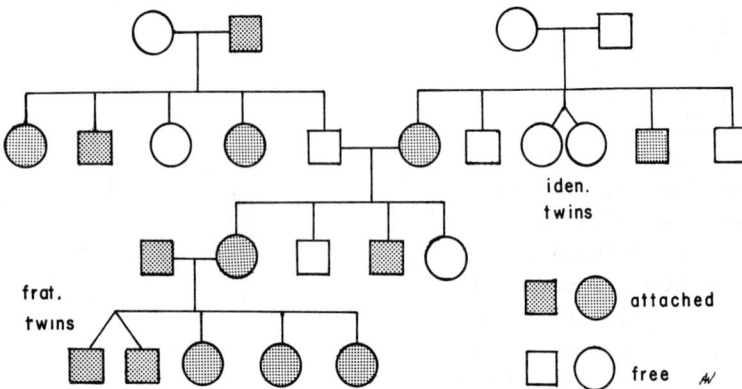

Fig. 1.5. *A family pedigree showing inheritance of ear lobes. Those with ear lobes attached directly to the head are shown by the shaded squares or circles and those with a free-hanging ear lobe are shown in the unshaded squares and circles. The method of indicating identical and fraternal twins is also shown.*

proper shading of the squares and circles. A study of the pedigrees in Fig. 1.5. indicates that the characteristic of attached ear lobes results from a gene which is recessive to the one causing free-hanging ear lobes. Of course, much more extensive pedigrees would be necessary before any conclusions were drawn definitely.

Cytology. Cytology is a branch of biology which deals with a study of the details of cell structure. Since the physical basis of heredity lies within the cell, the geneticist must turn to cytology for the answers to many of his problems. Through cytological studies we can see rod-like bodies within the cells. These are especially clear-cut during, and just preceding, cell division. We call these bodies "chromosomes" and they are known to be the carriers of the genes. Hence, through the microscope we can view the physical processes of the assortment of the elements of heredity which are expressed in breeding experiments.

Many questions about the results obtained in experimental breeding have been solved through cytological observations. For example, we may consider the case of waltzing in mice. Waltzing in mice is due to a defect of the inner ear which makes it difficult for them to keep their equilibrium; the mice, thus, sway from side to side in a manner suggestive of waltzing. This characteristic is

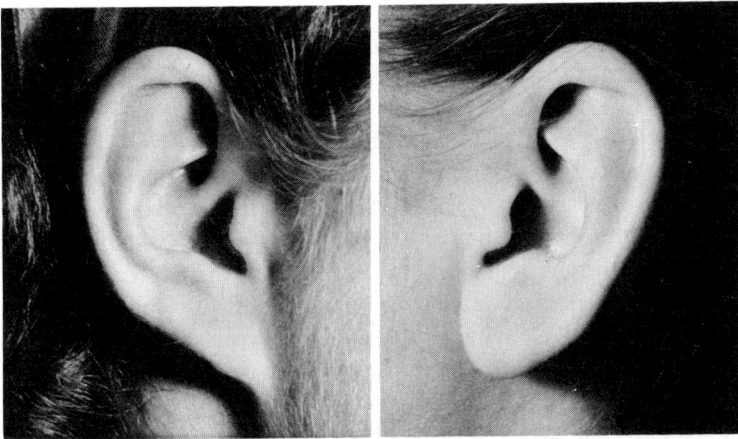

Fig. 1.6. Photographs showing the difference between an attached ear lobe (left) and a free-hanging ear lobe (right).

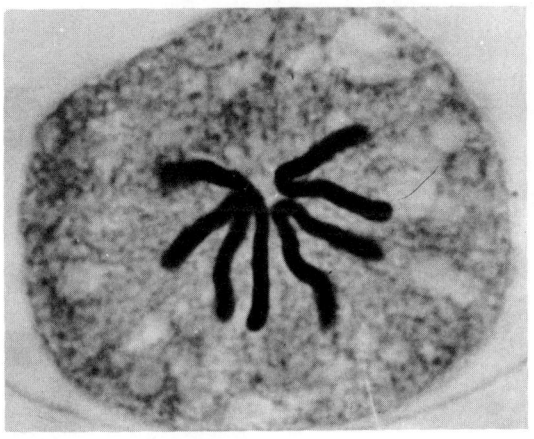

Fig. 1.7. Chromosomes, the carriers of the genes of heredity. These are typical chromosomes as seen when a cell is in process of preparation for division. This cell is from a parasitic worm, known as Ascaris, which has only four chromosomes in each of its body cells.

known to be due to a recessive gene and normally there must be two of these genes before the characteristic is expressed—if there is a dominant gene for normal ears the recessive gene will not have its effect. In one cross between a pure line normal and a waltzing mouse, however, a waltzing offspring appeared. This should not have occurred—the dominant normal gene should have suppressed the recessive waltzing gene. Cytological studies have given us the answer—the studies showed that a part of a chromosome carrying the normal gene had been broken off and, hence, there was no inhibition of the recessive gene which causes the waltzing trait.

Biochemical Genetic Study. In recent years there has been an increasing interest in the biochemical reactions associated with gene action. In the early years of genetic study, the emphasis was placed on learning the ultimate effects of gene action—a certain gene was present and a certain trait resulted. As this information has accumulated, there has come a desire to know how the effects are produced. Through what physiological mechanisms do the genes operate?

As an example of this type of study, let us consider *albinism* in

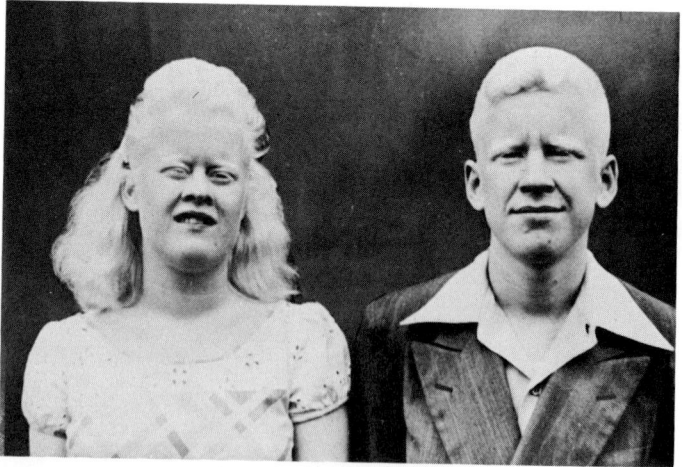

Fig. 1.8. Albino sister and brother. Genes inherited by these young people prevent the proper production of the pigment melanin which gives color to the hair, skin, and iris of the eyes. Through physiological genetics we are learning what takes place in the body chemistry which prevents the normal formation of the pigment.

man. This trait is characterized by an inability to properly synthesize the pigment melanin which gives color to the hair, skin, and iris of the eyes. Hence, the skin is unpigmented, the hair is almost white or straw-colored, and the eyes are usually pink because of the blood showing in the unpigmented irises. Pedigree studies have shown that the condition is usually due to a recessive gene—when a person receives two of these genes he is an albino. But why?—how does the albino's body differ in the physiological reactions from those of a person who has normal pigmentation? Biochemical geneticists have found that these genes fail to produce an enzyme which is a necessary part of a chain of reactions in the breakdown of an amino acid, a breakdown which eventually results in melanin production.

We have learned much about the biochemistry of genetics through studies of small, relatively simple forms of life, such as molds and bacteria. It is not difficult to control and analyze the chemical environment of such organisms and we can thus learn the part played by the genes in the various reactions associated with metabolism. Discoveries of the way in which genes (through enzymes in a mold) govern chains of reactions laid the foundation for the study of the chain of reactions which results in melanin production in man.

2: EARLY DEVELOPMENT OF GENETIC KNOWLEDGE

Genetics is a comparatively young science—it has been developed mostly during the present century, but it is interesting to go back into earlier studies and speculations to see how the background for modern genetics has evolved.

EVIDENCES OF ANCIENT UNDERSTANDING OF GENETICS

Man's application of genetic principles in the breeding and selection of cultivated plants and domestic animals dates back to ancient records. For example, early Chinese accounts show that

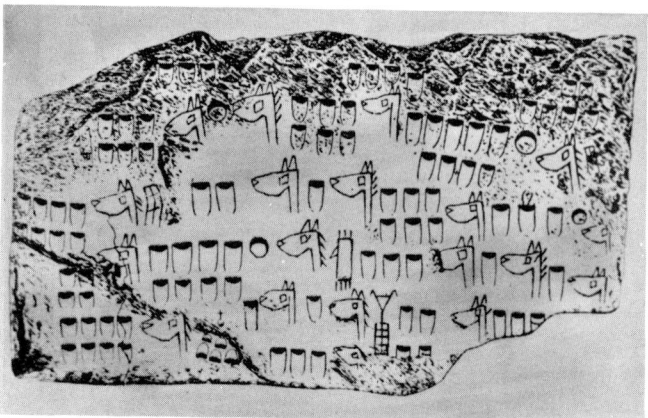

Fig. 2.1. *A pedigree of horses six thousand years old. This ancient stone tablet excavated in Chaldea shows inheritance in five generations of horses. The symbols shown here include three types of mane (erect, pendant, and maneless) and three types of profile (convex, straight, and concave). (From Amschler,* Journal of Heredity.)

superior varieties of rice had been developed almost 6,000 years ago. Likewise, a Babylonian stone tablet dating back to about 4000 B.C. gives the pedigree of five generations of horses showing how characteristics of the head and mane were transmitted. Ancient Egyptian paintings depict pictures of men cross-pollinating the date palm. The great number and variety of breeds of dogs that exist today show that our distant ancestors were using genetic breeding techniques to produce animals best suited to their needs.

THE GREEK PHILOSOPHERS

The earliest recorded speculations about the nature of the transmission of inherited characteristics come from the Greek philosophers—Pythagoras, Empedocles, and Aristotle.

Pythagoras. Pythagoras, who lived about five centuries before Christ, proposed the concept that moist vapors descend from an animal's body during coitus, and that these form body parts in the embryo similar to the body parts from which they came.

Empedocles. Another philosopher of the same period, Empedocles, proposed that each parent produces a semen in the various body parts and that these unite and form similar parts in the embryo. He suggested that not all of the semen from both parents is used and, hence, a child will show some characteristics of one parent and some of another while some characteristics of the parents will be lacking altogether.

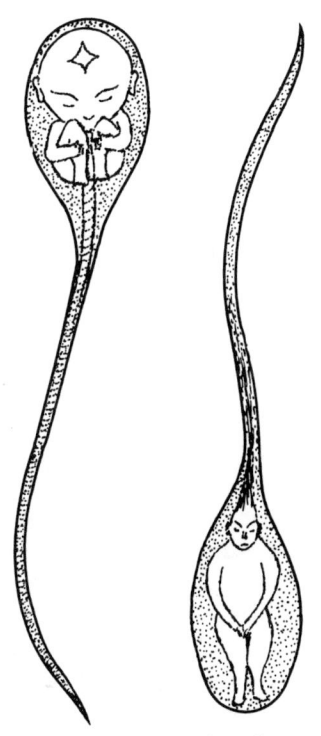

Fig. 2.2. *Preformed embryos within sperms. Some early investigators imagined that they could see tiny human embryos within the human sperms. These are copies of drawings showing what two men imagined they saw. (Left, after Hartsoeker, 1694. Right, after Dalempatius, 1699.)*

Aristotle. The great Aristotle, some two hundred years later, modified this theory somewhat by suggesting that blood is the true element of heredity and that there is a continuous flow of blood directly from the parents to their offspring down through the generations. The semen of the male was taken to be highly purified blood, while the "semen" of the female (confused with the menstrual fluid) was taken to be less highly purified. This viewpoint carried great weight with scholars for many centuries and even today there are some who feel that blood has a hereditary influence, as evidenced by the reluctance of certain people to accept blood transfusions from persons of another race. We use the terms, "blood relative," "blue blood," "bad blood," and "blood line," evidence of how Aristotle's concept has influenced the thinking on inheritance. As late as the seventeenth century there were illustrations in medical books showing the stages of the supposed coagulation of an embryo from the "semen" of the parents.

THE DAWN OF MODERN CONCEPTS

The seventeenth and eighteenth centuries proved to be profitable years for the advancement of genetic concepts. Scientists who conducted experiments and advanced their own theories include: William Harvey, Anton van Leeuwenhoek, Jan Swammerdam, Regnier de Graaf, Pierre-Louis de Maupertuis, Charles Bonnet, and Kaspar Friedrich Wolff.

William Harvey. It was during the seventeenth century that William Harvey (1578–1657) in England decided to test the theory of Aristotle which had stood for so long. He mated twelve female deer and killed some of them during various stages of pregnancy. He never found anything like the coagulating fluids in the uterus. Instead, he saw an embryo first in a deer killed several weeks after mating, and this was a very small embryo that did not even look like a deer. In later examinations he found that as it grew larger the embryo gradually assumed the appearance of a deer.

Leeuwenhoek. During the latter part of the seventeenth century a great impetus to this study was brought about by the development of the microscope. A Dutch microscope-maker, Anton van Leeuwenhoek (1632–1723), saw living sperms in the semen of a number of different animals, including man. He noted the association of these sperms with the eggs of frogs and fishes and reasoned that it is a union of these two which results in the formation of an embryo.

Swammerdam and the Preformation Theory. Another Dutch scientist, Jan Swammerdam (1637–1680), who continued the microscopic study of sperms, imagined that he could see a tiny embryo within the head of a human sperm. Using this mistaken observation as a basis for further work, he proposed the preformation theory, which held that the sperm furnishes the embryo while the uterus of the female is necessary as a site for the enlargement of this embryo.

Graaf and the Discovery of Mammal Eggs. Still another Dutch scientist, Regnier de Graaf (1641–1673), studied the ovaries of

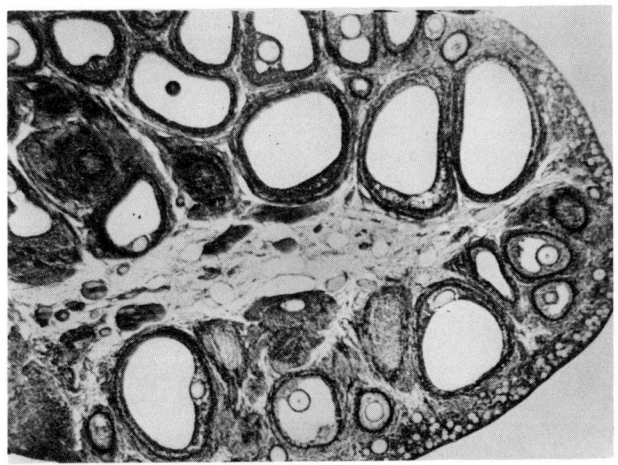

Fig. 2.3. Section of a portion of a human ovary showing eggs contained within Graafian follicles. These are so named because de Graaf first saw these and recognized the fact that mammals produce eggs in a manner very similar to birds.

different animals and found that the ovaries of mammals produce eggs just the same as ovaries of birds although the mammal eggs are much smaller. He proposed that the egg of a mammal breaks from the ovary, is fertilized by union with a sperm, and makes its way to the uterus where it develops into the embryo. He found cases of extrauterine gestation which proved that there is nothing about the uterus which is necessary for embryonic development.

These were sound observations and conclusions, but they were clouded by speculations that it is the egg and not the sperm which contains the supposed preformed embryo.

Maupertuis. A Frenchman, Pierre-Louis de Maupertuis (1698–1759), took issue with the idea of preformation and asserted that the plain facts of biparental inheritance rule out the possibility that either the sperm or the egg furnishes a tiny embryo. He performed many breeding experiments and collected the family pedigree of human beings. From his studies he proposed the theory that tiny particles migrate from all parts of the body to the reproductive organs where they form the semen. The particles were supposed to retain some sort of "recollection" of the body areas from whence they had come and to reproduce similar body structures in the embryo. Furthermore, he assumed that a particle from one parent might dominate over a particle from the other and thus

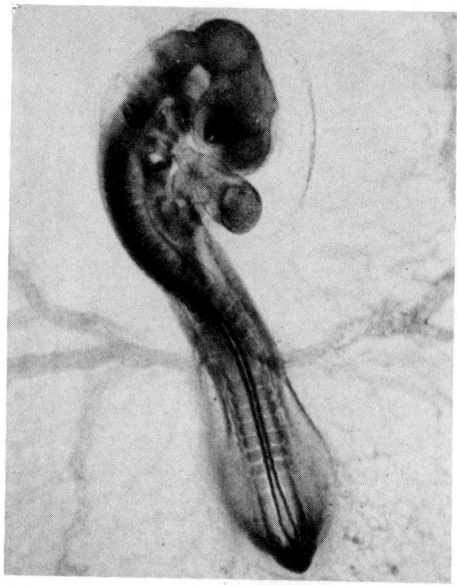

Fig. 2.4. A chick embryo 48 hours after incubation was begun. It was through a study of the development of the chick embryo that Friedrich Wolff discovered that body organs develop in a logical sequence from the fertilized egg, and that there is no preformed embryo.

account for the fact that a child would show some characteristics of one parent and some of the other.

Bonnet and the Encasement Theory. Another theory which received much support was proposed by a Swiss scientist, Charles Bonnet (1720–1793). His encasement theory held that every female animal carries within her body the prototypes of all of the creatures which will ever descend from her, one generation within the other somewhat like a series of Chinese boxes. He did not believe these to be preformed embryos, but assumed that particles contained therein have the power to form embryos. He concluded that each form of life is created with a limited number of generations within the body and that when these have been exhausted the species becomes extinct.

Wolff and Epigenesis. The German scientist, Kaspar Friedrich Wolff (1733–1794), made extensive studies of the development of the chick embryo. He proved that not only was there no preformed embryo, but that, instead, the body organs developed in a logical sequence. He proposed the theory of epigenesis which held that both male and female gametes (sperms and eggs) contain bodies which, after fertilization, become organized into body organs. This is quite similar to our present concept of the genes as the basis of inheritance.

NINETEENTH-CENTURY CONTROVERSIES

Nineteenth-century controversies were concerned with the concepts of use and disuse, the inheritance of acquired characteristics (upon which much reliance has been placed by Russian workers in the twentieth century), natural selection, pangenesis, the germ plasm theory, and the mutation theory.

Use and Disuse. A Frenchman, Jean Baptiste Lamarck (1744–1829), proposed a theory which has had great impact on the study of heredity even to the present day. He emphasized the importance of "use and disuse" in the establishment of inherited characteristics. According to his interpretation of this principle, body organs and structures which are used extensively become more highly developed in the individuals that so use them. The descendants will inherit these adaptations and through still further use will make them even more highly developed, and so this process will continue down through the generations. As an example of this,

Lamarck theorized that wading birds have long necks because each successive generation of these birds has stretched the neck more and more as they have reached down into the water for food. On the other hand, organs which are not used tend to degenerate and the weakened organ is also passed on to the offspring. As an example of this, he cited the fact that the ant bear eventually lost its teeth because it had developed the habit of swallowing its food whole. Today we recognize the importance of use and disuse in influencing the establishment of characteristics in races, but we have more satisfactory explanations of the mechanism involved.

Inheritance of Acquired Characteristics. Lamarck's theories brought into focus the concept of inheritance of acquired characteristics. He even went so far as to state that a race of one-eyed people could be developed if the left eye were to be removed from a group of children of each generation. He maintained that an animal's needs determine its desires, its desires determine the use or disuse of its body parts, use or disuse brings about modifications, and these modifications are inherited.

The Russian Viewpoint toward Acquired Characteristics. It is interesting to note at this point that, while reputable geneticists throughout the world have rejected Larmarck's theory of inheritance of acquired characteristics, there has been a comparatively recent resurgence of the belief in Russia. An obscure plant breeder, Trofim D. Lysenko, did some experiments on wheat and tomatoes during the 1930's. He believed that he had found evidence that the plants could be improved genetically through improved growing conditions. Although he had had no fundamental training in genetics, he believed his discovery could be applied to all forms of life. Such unsubstantiated proposals are likely to be made in any country, but usually they are not taken seriously. Unfortunately for the trained geneticists of Russia, however, Lysenko convinced the Soviet authorities that his was the correct viewpoint and that the real geneticists were enemies of the state. There resulted an elimination of the best geneticists of the country and Lysenko was put in power to revolutionize agricultural production by his methods. Several years ago, however, the Russian leaders began to realize that Lysenko's methods were not producing results and he was demoted to a role of lesser importance. The present regime in Russia is encouraging the redevelopment of modern genetics, but there has been a great loss to the country because of the Stalinist suppression of genetic research along modern lines.

Natural Selection. England's Charles Darwin (1809–1882) made observations and proposed theories that have had a profound influence on the development of modern genetic knowledge. On his voyage around the world in the *Beagle* he observed many forms of life and concluded from his data that no species was created in exactly the same form in which it exists today—all his evidence pointed to the fact that in each species there had been significant modifications in structure and function. After much study and deliberation he wrote his famous book, *The Origin of Species by means of Natural Selection* (1859), which propounded the basic principles of his theory of evolution of life on earth. First, there is *overproduction* of offspring in all forms of life. Second, there are *variations* among these offspring—variations due to heredity. Third, there is a *struggle for existence.* Fourth, there is a *survival of the fittest* and the most favorable variations become established. This is the way, according to his theory, that changes in nature take place and new species arise.

Pangenesis. As Darwin attempted to find an explanation for the variety of inherited characteristics which are present in all forms of life, he proposed the provisional hypothesis of pangenesis. This assumes that every part of the body produces minute pangenes which are carried to the reproductive cells. When the reproductive cells unite, the pangenes contained therein reproduce the body parts whence they came. This would allow for inheritance of acquired characteristics—for example, a blacksmith's well-developed arms would pass to his children through his pangenes. Darwin even tried to apply the hypothesis to plants in as much as he assumed that their pangenes went not only to pollen and ovules, but also to the buds and twigs which had the power to produce entire plants asexually.

The Germ Plasm Theory. During the latter part of the nineteenth century the German biologist, August Weismann (1834–1914), began to question the theories of Darwin and Lamarck. He noted the life cycle of one-celled animals which divide by fission with an indefinite continuity of their protoplasm. Believing that something of this nature must be true of the reproductive tissue of higher animals, Weismann formulated the germ plasm theory. This theory held that (1) there is a special germ plasm in all

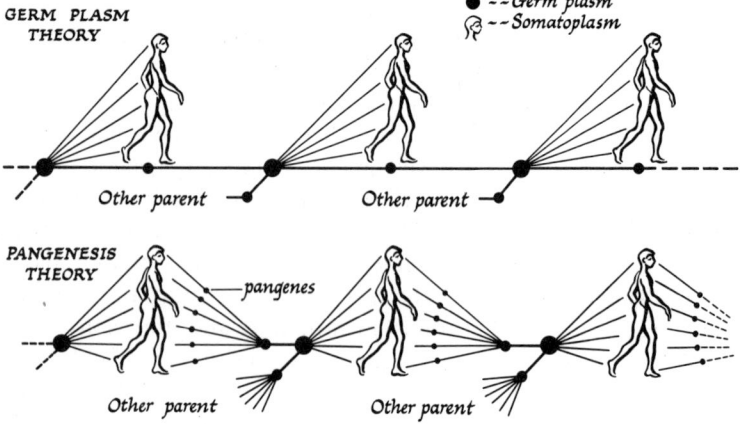

Fig. 2.5. The germ plasm theory of Weismann contrasted with the pangenesis theory of Darwin. Weismann pictured the germ plasm as a continuous stream producing somatoplasm as it flowed through the generations. Darwin thought of each generation as producing the germ plasm for the next generation. (From Winchester, Genetics, Houghton Mifflin).

multicellular organisms which preserves itself by repeated duplication and that (2) these cells give off cells which grow into the main part of the body, the somatoplasm, but that (3) the germ plasm remains isolated and is in no way influenced by the body tissues. Thus, the germ plasm forms a continuous stream flowing down through the generations, uniting with other germ plasm and forming somatoplasm along the way, but itself remaining unaltered.

Weismann conducted many experiments to show the superiority of his theory over those theories that dealt with the inheritance of acquired characteristics. His most famous experiment consisted in cutting off the tails of a group of mice for 22 generations. When the tails were allowed to grow from the mice of the next (23rd) generation they were found to be as long and as normal as if there had been no mutilation of previous generations. Chromosomes within cells were discovered about this time and Weismann was one of the first to believe that the chromosomes are the bearers of the units of heredity.

The Mutation Theory. A Dutchman educated in Germany, Hugo DeVries (1848–1935), was the first person to note the importance of mutations—spontaneous changes in heredity. Darwin had noticed the appearance of occasional new types among the different organisms and had called them "sports," but had failed to realize that these were the source of the variety in heredity which he had sought to explain by other methods. DeVries noted unusual forms of evening primroses growing among the more common types in the meadows. The variations included dwarf plants, flowers with double petals, plants with unusual leaf types, and many more. He found that these variant types could be transplanted and seeds taken from them would produce more plants like those from which they came. Consequently he formulated his mutation theory—that living organisms occasionally, but regularly, produce new types of offspring through sudden changes (mutations) in the hereditary mechanism. Scientists have since learned that the new types of evening primroses were the result of chromosome changes and not of gene mutations as the word is used today, but DeVries' theory remains essentially correct.

GREGOR MENDEL AND THE DEVELOPMENT OF THE GENE CONCEPT

In the midst of the controversies aroused by the theories of Darwin, Weismann, and others, a new concept was being developed by the Moravian monk, Gregor Mendel (1822–1884), who was quietly conducting breeding experiments with the garden pea. Mendel eventually discovered the true secrets of genetic inheritance and became known as the "father of modern genetics."

In his youth Mendel joined the Augustinian monks in the monastery at Brünn. Later he studied mathematics and natural history at Vienna, where he became interested in the hybridization of plants. When he returned to Brünn he chose garden peas as a subject for his experiments because (1) they could be obtained in a number of pure-breeding varieties, (2) they were normally self-pollinating, and (3) the hybrids between two varieties were fully fertile. He obtained 22 varieties for his experiments and used artificial cross-pollination to obtain his results. After eight years of careful breeding and analysis of his results he worked out the method of gene transmission in an exact manner. (Details of some of his findings are given in Chapters 4 and 5.) But when he presented the results of his work (in 1866 and 1869), the scientific world, engrossed in other radical new theories, failed to realize that here were the answers to many of the questions raised by Darwin. So Mendel died without recognition.

THE EARLY TWENTIETH CENTURY

In the early years of the present century several investigators, including DeVries, in checking over the literature in order to find an explanation for their own experiments, came across the paper by Mendel and recognized its monumental importance. This was the dawn of modern genetics. Many investigators took up the work, the foundations of which had been laid by Mendel.

In Denmark, Wilhelm Johannsen developed his *pure-line concept* as a result of his experiments on beans. His work suggested that selection for size of beans was effective for one generation only, after which the lines bred true regardless of selection. There was variation in each line, but this variation was constant no matter whether large or small beans from the pure line were planted. He coined the word *gene* which has been adopted universally since that time.

In Sweden, H. Nilsson-Ehle worked on wheat genetics and achieved important applications of the science in the improvement of this field crop. Erwin Baur and Carl Correns in Germany found that not all hereditary units (genes) segregate independently as described by Mendel, but some tend to remain together for generation after generation. In England, W. Bateson and R. C. Punnett worked on sweet peas and developed the principle of **linked genes** as first discovered by Correns and Baur. In the United States, Columbia University became a center for early genetic investigations. E. B. Wilson did pioneer work on *cytogenetics* and made important discoveries on the relation of chromosomes to heredity. To Thomas Hunt Morgan, of Columbia University, goes the distinction of being the first to use the fruit fly, *Drosophila melanogaster*, as a form for experimental breeding.

From these beginnings the science of genetics has spread and is now represented by many geneticists in most of the countries of the world. We will learn about some of these as we progress with our study.

3: THE PHYSICAL BASIS OF HEREDITY

The physical basis of heredity is to be found in the structure and functions of living cells, more specifically in the genes, the chromosomes, and in the processes of cell division and maturation.

GENES AND CHROMOSOMES

Genes form the physical hereditary link between generations. Within the nucleus of the cells of all types of organisms there are many of these tiny constituents. A typical human body cell, according to the best available estimates, contains about 40,000 genes. If your eyes are blue, or your hair is curly, or you have artistic talent, it is because somewhere in this vast number of genes there are some genes that have made you this way. These genes in your body cells today are typically of the same number and kind as those that were present in the one original zygote (fertilized egg) from which you descended. It is evident, therefore, that there must be gene duplication before each cell division.

Gene Duplication. Living matter has the unique power of self-duplication—the ability to produce more matter like itself. Since genes serve as the directing force of living reactions, we can hope to understand these reactions better as we learn more about gene duplication. Much study has gone into this topic in recent years and the evidence indicates that each gene first splits in two, and then each half attracts unto itself all the necessary chemical matter and arranges it in such a way as to restore the original gene structure. (This process is discussed more fully in Chapter 14.)

Chromosome Duplication. After the genes within a cell are duplicated, there remains a problem which must be solved before the cell can divide. Each of the genes must be so assorted that the daughter cells will receive one of each of the partner's genes formed by gene duplication. With genes numbering high in the thousands, this would seem to be a major engineering feat. It appears greatly simplified, however, when we learn that genes do not exist as separate units, but are arranged in a linear order on thread-

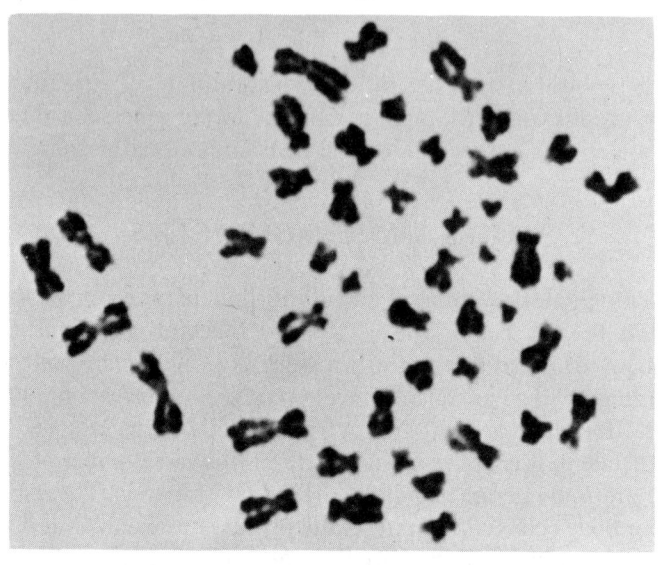

Fig. 3.1. *The 46 human chromosomes. In this cell the chromosomes are shortened and have already become double in preparation for cell division. The cell was crushed flat so all chromosomes would be in one plane and clearly seen under the microscope. This is one of the first photographs to establish the human chromosome number as 46. (Photo by Joe Hin Tjio, Estacion Experimental de Aula Dei, Spain).*

like bodies known as chromosomes. These become shortened and thickened during cell division and can be clearly seen under the microscope.

Each species of plant or animal tends to have a constant chromosome number. In man, for instance, the number found in typical body cells is 46. It is certainly much simpler mechanically for the cell to handle the duplication and proper segregation of 46 such bodies than it would be if there were about 40,000 individual units to be handled. The chromosome number of some common organisms is shown in the table below.

Organism	Diploid Chromosome Number
Drosophila melanogaster (fruit fly)	8
Garden pea	14
Onion	16
Corn	20
Opossum	22
Bullfrog	26
Honey bee	32
Domestic swine (hog)	38
Mouse	40
Man	46
Potato	48
Monkey (cebus)	54
Crayfish	200

MITOSIS

Mitosis is the term given to the process of cell division where there is a duplication of the genes, followed by a duplication of the chromosomes, a segregation of the chromosomes, and, finally, a splitting of the cell into two parts. The events of mitosis are most conveniently studied as a series of phases.

Interphase. This is the phase between mitoses. It is sometimes called the resting stage because there is no mitotic activity, but the cell may be at the height of its metabolic activity and the term "resting" may leave a false impression. In typical cells at this stage, the chromosomes are extended to a great length and may be so thin that they are not easily visible under the microscope. In this very extended state it is customary to speak of these chromosomes as **chromonemata.** Each of these chromonema (singular) bears a specialized body known as a **centromere.** This is located at some specific point and bears an important function in separating the chromosomes.

The chromonemata are all contained within the nucleus of the cell. If the cell is from a higher animal or one of the lower plants, there may be a body, known as the **centrosome,** in the cytoplasm outside the nucleus. This body consists of a central particle, the **centriole** (in some cells there are two centrioles), from which rays radiate out into the cytoplasm. These radiating rays form the **aster.**

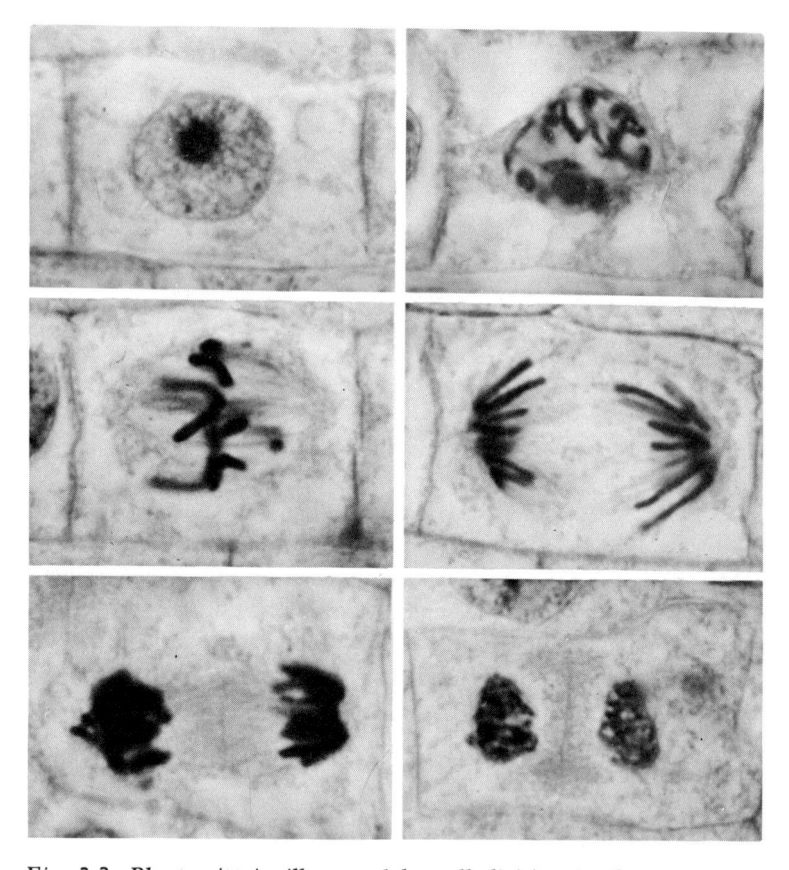

Fig. 3.2. Plant mitosis, illustrated by cell division in the onion root. Stages shown from left to right and top to bottom are: Interphase, prophase, metaphase, anaphase, early telophase, and late telophase. (From Winchester, Genetics Laboratory Manual, Wm. C. Brown Co.)

Prophase. In typical mitosis the genes become duplicated while the chromosomes are in a long, thin chromonema state. This marks the beginning of the events of mitosis. At this stage the chromonemata are somewhat coiled, the so-called *minor coils*. The chromonemata then become duplicated, but the centromeres do not duplicate at this time. Hence, there are two chromonemata for each centromere. During the early part of the prophase additional coils (the *major coils*) appear and this makes the chromosomes shorter and thicker. At the same time a *matrix* begins to be deposited around the double-coiled chromonemata. As the coiling and matrix deposition continue, the chromosome assumes the appearance of a rod-like or sausage-shaped body. By this time the matrix is so heavy that the coiled chromonemata are usually not visible with the commonly used microscopic techniques, but by special methods they can be revealed in many forms. During the latter part of the prophase the dual nature of the chromosomes may become apparent and we refer to the half-chromosomes as *chromatids.*

In the cytoplasm other changes are taking place. The centriole divides (if it is not already double) and the two centrioles move apart. Then between them appear lines which form the *spindle figure.* The nuclear membrane gradually disappears and the chromosomes become oriented in the center of the spindle figure. In forms that do not have a centrosome, however, the spindle figure makes its appearance without the preliminary steps mentioned. The lines between the opposite poles of the spindle figure are called *spindle fibers,* but evidence indicates that they are not fibers in the true sense of the word, but rather lines of force brought about probably by the rearrangement of protein chains of molecules into a longitudinal sequence.

Metaphase. When the chromosomes have moved to the central portion of the spindle and become arranged in an *equatorial plate* equidistant from the two poles of the spindle, the metaphase begins. If the chromosomes are long they may extend from this plate, but the centromeres seem to be anchored at this region. A spindle fiber appears to be attached to each centromere. The centromeres now become duplicated and begin moving to opposite poles, apparently being guided on their way by the spindle fibers. This movement pulls the two chromatids of each chromosome apart so that, as they become separated, each chromatid becomes a chromosome. Since the spindle fibers are not actual fibers, we know that they cannot "pull" on the centromeres as a thread would pull. Rather we might think of them as guiding lines that orient the centromeres as they move to the poles.

Anaphase. This phase begins as soon as the chromatids have separated and continues until the two groups of chromosomes have reached the poles of the spindle figure. The centromere leads the way to the poles and the rest of the chromosome appears to be "dragged" along behind. Thus, if the centromere is terminal, the

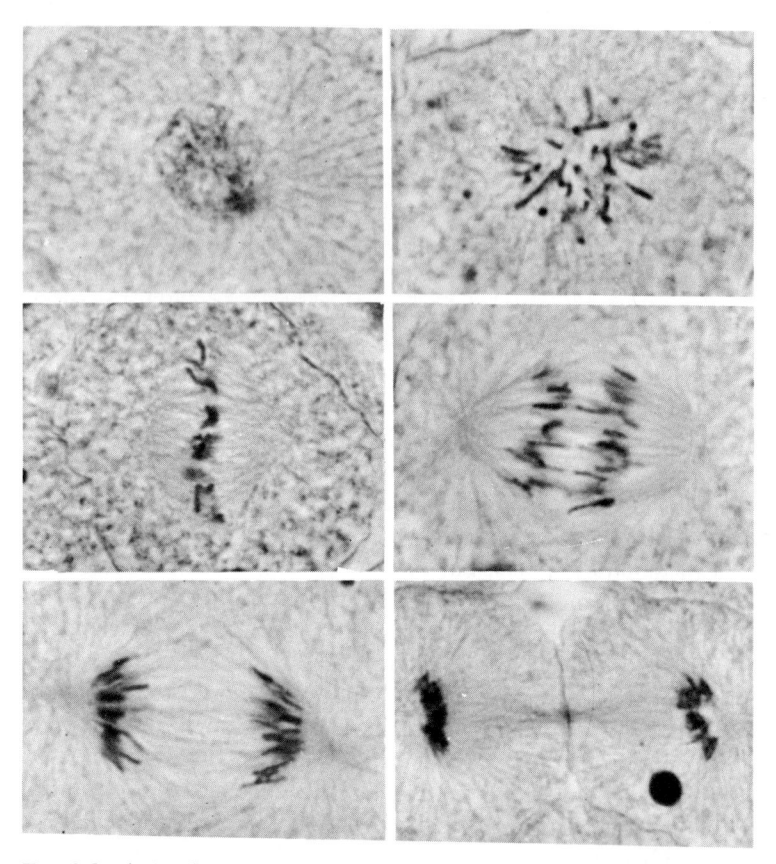

Fig. 3.3. *Animal mitosis as illustrated by cell division in the whitefish. Stages shown from left to right and top to bottom are: Interphase, prophase, metaphase, early anaphase, late anaphase, and telophase.* (From Winchester, Genetics Laboratory Manual, Wm. C. Brown Co.)

chromosome will be in the form of a straight rod. A central centromere, however, will yield a V-shaped chromosome during this phase, and a sub-terminal attachment will result in a J-shape. The centromere follows a spindle fiber to the pole as described.

Telophase. During this phase there is a reconstitution of the nucleus of the interphase. After the chromosomes reach the poles they tend to become longer and thinner as the chromonemata lose their major coils and the matrix is gradually dissolved. At the same time, the actual division of the cell is taking place. In typical

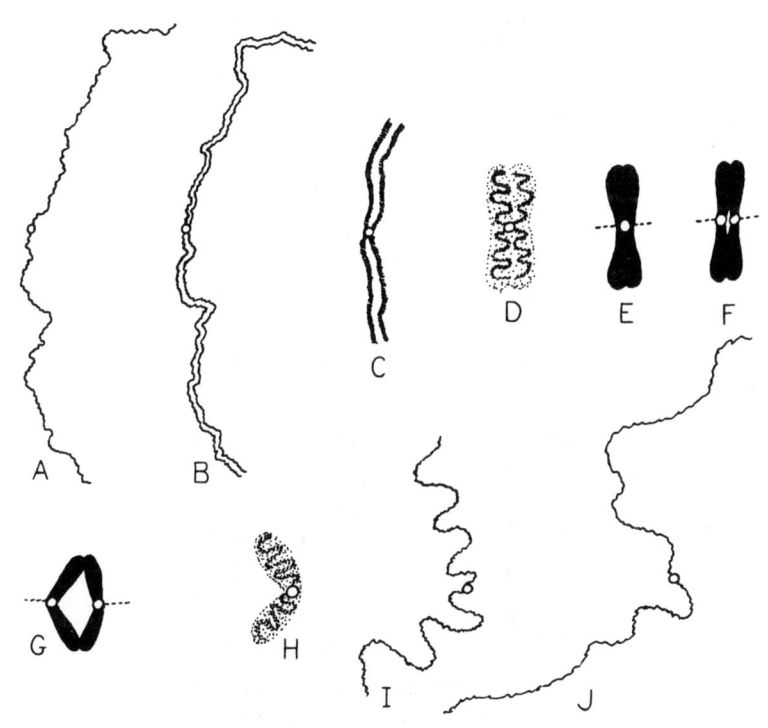

Fig. 3.4. Life cycle of a chromosome. A. *Chromonema in interphase.* B. *Chromonema duplicates and forms two.* C. *Chromonemata shorten and thicken by coiling in early prophase.* D. *Shortening and thickening continues with more coiling and matrix is deposited around them.* E. *As matrix becomes heavy we see two chromatids making up this prophase chromosome.* F. *Centromere becomes duplicated during metaphase.* G. *Chromatids separate and form two chromosomes at beginning of anaphase.* H. *(one chromosome omitted) Matrix begins to fade and chromonema becomes visible in telophase.* I. *and* J. *Chromonema becomes extended in the interphase condition.*

plant cells, a **cell plate** forms between the two daughter nuclei, becomes a cell wall, and cell division is accomplished. In most animal cells, on the other hand, a **cleavage furrow** develops on the outside of the cell. This gradually constricts and pinches the cell in two.

Daughter Cells and Twins. When the telophase has been completed, the two cells which have been formed are in their interphase. Through the process of mitosis each of these cells has received the same number and kind of genes present in the one

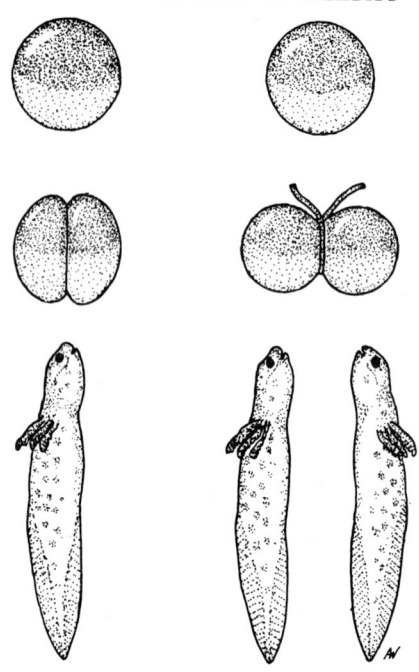

Fig. 3.5. *Artificial production of twins in the salamander. The young salamander embryo, in the two-cell stage, can be pinched in two by a piece of string and each cell forms a complete salamander (identical twins). Had the cells been allowed to remain together only one salamander would have resulted. This shows that the genes are duplicated in mitosis.* (From Winchester, Biology and Its Relation to Mankind, Van Nostrand).

cell from which it descended. This can be demonstrated rather dramatically by a simple experiment. Assume that the cell which will divide is the fertilized egg (zygote) of a salamander. In normal development each of these cells will then form one half of the body of a salamander through repeated divisions. However, in our experiment the two daughter cells can be separated from one another by pinching them apart with a string tied around them. In this case each cell will form a complete salamander. These will

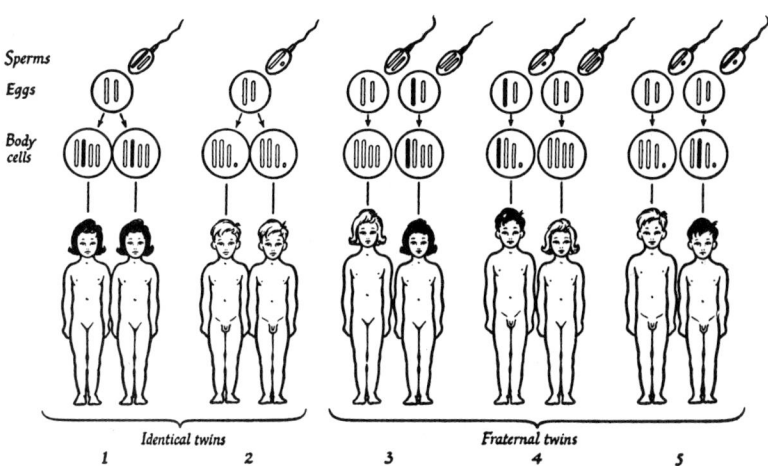

Fig. 3.6. How the two kinds of human twins are formed. When the embryo from one fertilized egg splits in two, identical twins are formed. These have identical heredity. Fraternal twins result when two different eggs are fertilized by two different sperms and may differ in inherited traits and sex. (From Clyde Keeler).

each have identical genes and, hence, identical heredity. Thus, we have evidence that mitosis must have achieved a true duplication of the genetic material of the zygote.

In human beings we often have a similar situation when there is a separation of cells of the early embryo and **identical (monozygotic) twins** result. These individuals are valuable sources of data for genetic studies because, since they have the same heredity, whatever differences such twins manifest must be due to environment. Of frequent occurrence among human offspring are **fraternal (dizygotic) twins** which develop from two separate fertilized eggs. Such twins are no more alike than brothers and sisters born at different times and may even be of different sexes. These individuals are also valuable sources of data, particularly for comparison with the monozygotic twins, for they will differ with respect to some genes and will show differences due to heredity as well as environment.

MEIOSIS

In all multicellular plants and animals which have sexual reproduction there is a special kind of cell division, known as meiosis, which achieves a reduction in the number of genes and chromosomes within the cell to one half. Otherwise, with the fusion of gametes, each carrying a full complement of chromosomes, there would be a doubling of chromosomes each generation.

In meiosis there are two cell divisions for only one gene and chromosome duplication. This results in the formation of four cells, each with only one half the number of genes and chromosomes present in the original cell. We speak of this reduced chromosome number as the *haploid number*, while the number of chromosomes in the cell before meiosis is the *diploid* number.

In animals the reproductive cells are formed directly from the haploid cells after meiosis, but in plants there may be some further cell division before gamete formation. We will consider the formation of animal gametes first—this is known as the maturation of the germ cells.

MATURATION OF ANIMAL GERM CELLS

The processes included in the maturation of animal germ cells are spermatogenesis (sperm formation), and oögenesis (egg formation).

Spermatogenesis (Sperm Formation). This process is similar in all animals; we will use man as an example. Man's body cells typically contain 46 chromosomes—the diploid number. Within the testes are tiny tubules which form the reproductive cells (sperms). If we cut one of these tubules in cross-section we can see that there is an outer germinal epithelium made up of cells known as *spermatogonia*. These are the cells which produce the sperms. Each of them contains the diploid number of chromosomes (46). As new cells are formed by mitosis, some cells begin to migrate toward the center of the tube and become *primary spermatocytes*. These will now divide by meiosis. We will examine one such cell.

As the cell goes into the prophase of the first division of meiosis, we notice an important difference from the prophase of a typical mitotic division. The chromosomes begin to pair, and they undergo *synapsis*. Within a cell each chromosome has a homologous mate which is of the same size and structure. (There is one pair in the male which are not perfect mates; we shall consider these

MATURATION OF ANIMAL GERM CELLS 33

SPERMATOGENESIS   OÖGENESIS

Primary Spermatocyte   Primary Oöcyte

MEIOSIS first division metaphase

Secondary Spermatocytes   First polar body   Secondary Oöcyte

MEIOSIS second division metaphase

Spermatids

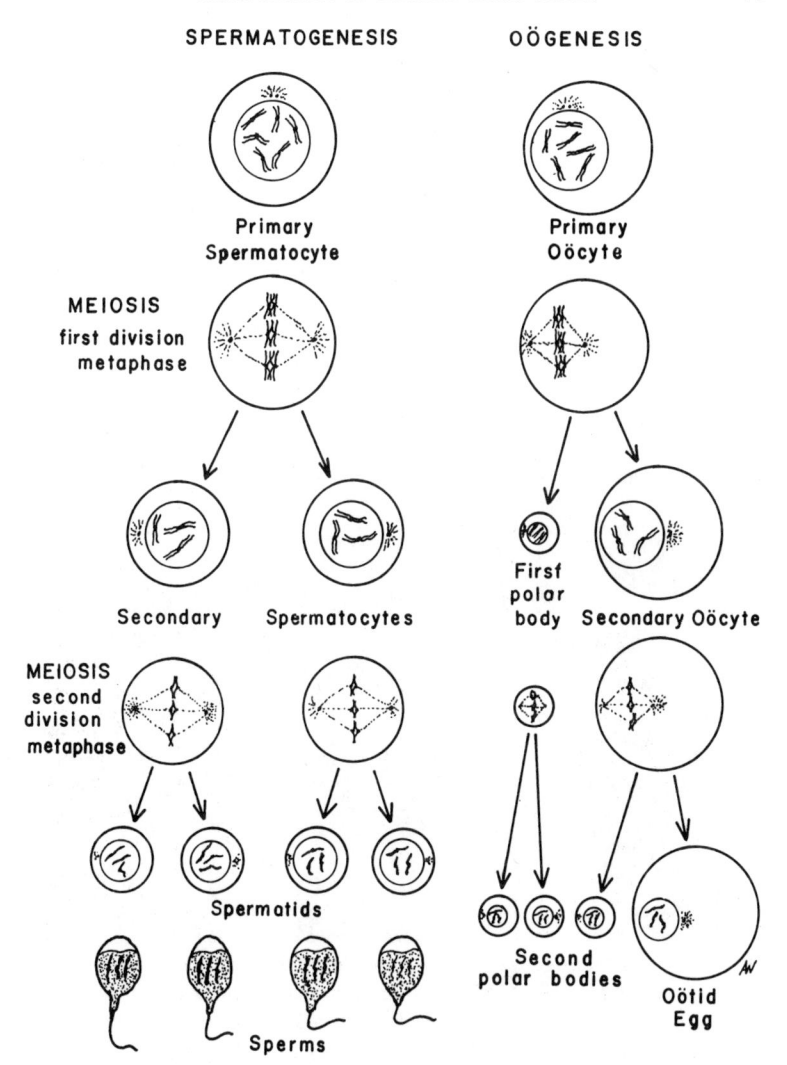

Sperms   Second polar bodies   Oötid Egg

Fig. 3.7. Diagram of sperm and egg formation by meiosis. Only six chromosomes are shown, but there would be 46 in human diploid cells.

further in our discussion of sex determination.) Some attraction seems to come about during this stage which causes the partner chromosomes to synapse. In mitosis such an attraction is generally not present and there is no such pairing of the chromosomes. The pairing in meiosis occurs before there is any visible evidence of

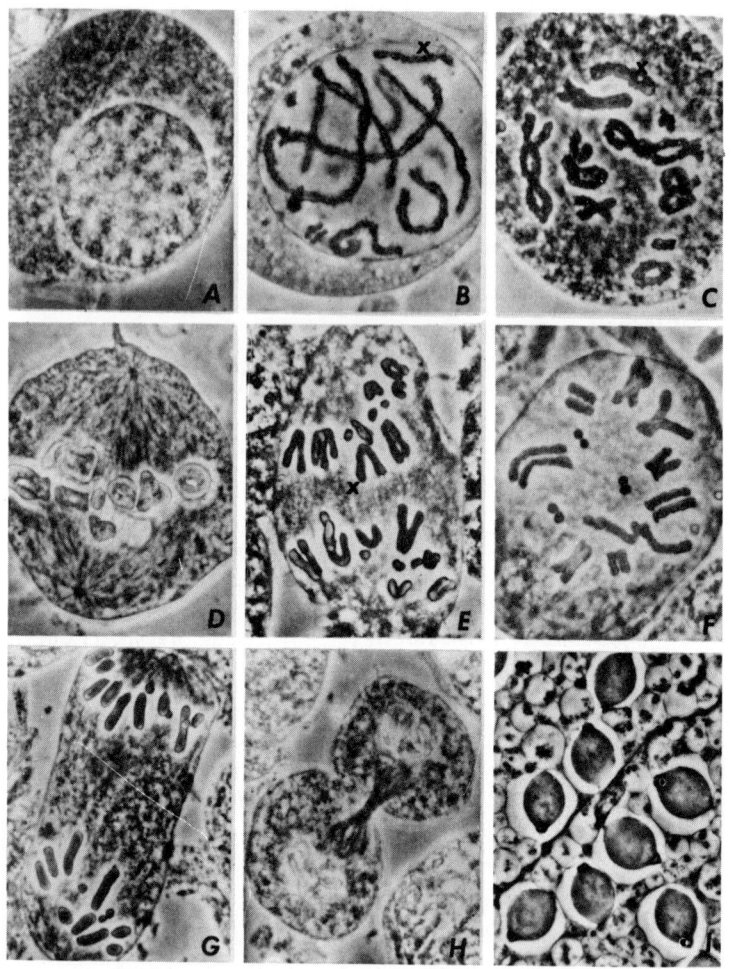

Fig. 3.8. *Spermatogenesis in the grasshopper.* A. Primary spermatocyte in interphase. B. Early prophase showing paired chromosomes. C. Late prophase, chromosomes have shortened and thickened and formed chromatids. The chromosomes are paired and each chromosome is formed of two chromatids, hence there are four chromatids to each chromosome pair. D. Metaphase. E. Anaphase; the chromosomes have separated and are moving to the poles. The double nature of each chromosome is apparent because the chromatids are somewhat separated except at the centromere end. F. Prophase; second division of meiosis. The two chromatids which form each chromosome are clearly visible. G. Anaphase; each chromosome now consists of only one part. H. Telophase. I. Spermatids, which will now form sperm without further division. These are photographs of living cells taken from the grasshopper testes without staining.

the duplication of the chromonemata, but after pairing it is clearly evident in many forms that the chromonemata of each chromosome is duplicated.

In the late prophase these paired, double chromosomes can be seen as four chromatids grouped together to form **tetrads**. These tetrads go into a typical metaphase plate, but there is no centromere division at this stage and the two centromeres already present separate and pull the chromosomes toward the poles. Thus, in the anaphase each chromosome is composed of two chromatids; such double chromosomes are called *dyads*. Then follows the telophase and the cell divides. Since each daughter cell will have 23 chromosomes, the reduction in chromosome number has been achieved, but the chromosomes are double and there is a second division of meiosis which permits separation of these into single chromosomes.

The two cells formed by the first division of meiosis are known as **secondary spermatocytes**. These go into the prophase of the second division, but there is no chromonemata duplication this time—they are already double. There is centromere duplication in the metaphase, however, and the dyads are pulled apart into single chromosomes. The telophase and the cell division follow. There are now four cells, **spermatids**, with the haploid chromosome number. Each of these forms a sperm through a condensation of nuclear material into a head, the formation of a middle piece which contains a small amount of food, and the development of a tail which serves for locomotion.

Oögenesis (Egg Formation). The ovaries of a woman contain a germinal epithelium which bears **oögonia**. These cells, of course, have the diploid number of chromosomes (46). Some of them form the **primary oöcytes** which go through meiosis in a manner similar to that described for spermatogenesis. There is one fundamental difference, however, for there is an unequal distribution of the cytoplasm in each division which results in one large and one small cell in each case. The spindle figure is formed near the edge of the primary oöcyte rather than in the center and only a very small amount of cytoplasm goes into one of the two cells formed. This results in a **secondary oöcyte** almost as large as the primary oöcyte and a very small **polar body**. The same is true of the next division of the secondary oöcyte, so that we obtain an **oötid** almost as large as the primary oöcyte. The first polar body also divides again and we end up with three of these polar bodies. These degenerate and thus play no part in reproduction.

FERTILIZATION

When the haploid sperm and haploid egg unite in the process known as fertilization, the diploid condition is restored and a new generation begins. The egg is always much larger than the sperm—it must have a storage of food to supply the developing embryo until the embryo can get nourishment from some other source. In mammals this means only until it can form an attachment to the uterus of the female. Hence, mammal eggs are relatively small. A human egg is only about the size of the period at the end of this sentence. In other animals, however, the stored

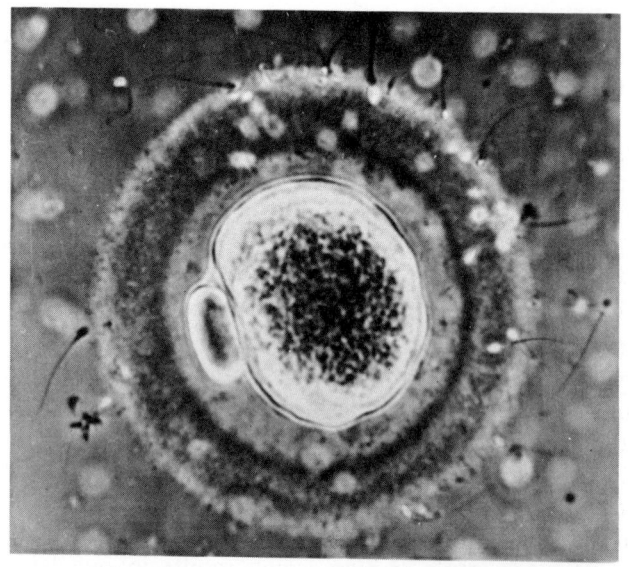

Fig. 3.9. Fertilization of a human egg. This photograph shows a living egg cell surrounded by its corona of follicle cells. Human sperms have been released around it and they can be seen attacking the corona. Some of the sperms have accomplished partial penetration of this barrier and are approaching the egg. One of the polar bodies given off by the egg is shown on the left. (Photo by Landrum B. Shettles).

food in the egg must supply the embryo until it can feed or be fed through its mouth. Thus, the eggs are generally much larger than mammal eggs. Birds, especially, have very large eggs.

Regardless of this difference in size, however, inheritance is about the same from both parents because both contribute about the same number of genes. If cytoplasm were an important factor in heredity we would expect inheritance to be much greater from the maternal side.

MEIOSIS AND GENE ASSORTMENT

As we have followed the process of meiosis it is evident that only one half of an animal's genes are carried by the sperm or egg. But, since the placement of chromosomes on the metaphase plate is due to chance, the different gametes formed will have different assortments of chromosomes. This process of assortment which takes place during the metaphase of the first division of meiosis might be compared to the shuffling of cards. It makes possible the great variety which may appear in the children of the same parents.

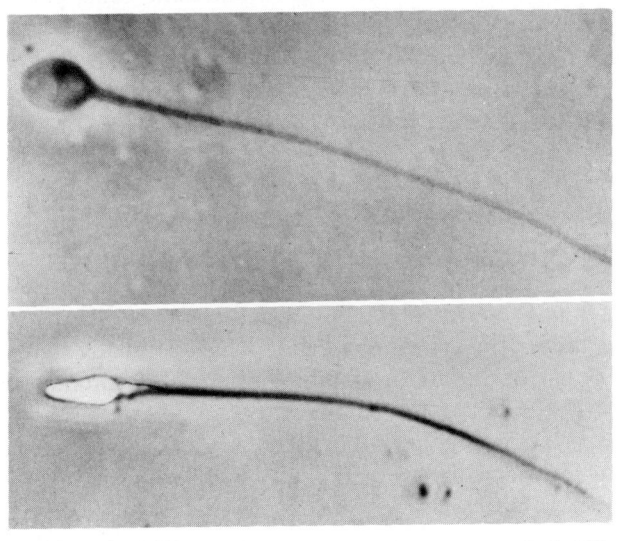

Fig. 3.10. Living human sperms, highly magnified. The top photograph shows the sperm head flat, while the lower photograph shows it from a side view. All of the hereditary potential that links a father to his child is contained within the head of the sperm.

If we omit identical twins from immediate consideration, we find that the chance of two children receiving the same assortment of genes is so small as to be considered impossible.

When meiosis occurs in the children of a couple and grandchildren are born, we find that each grandchild will be expected to receive about one fourth of its genes from each of the four grandparents. Here, however, the number is subject to considerable flexibility. While a child receives exactly half of its chromosomes from each parent, there is an independent assortment during the meiosis in the child; consequently, the number of chromosomes from the father (paternal) and the number from the mother (maternal) that go into any one gamete will vary according to chance arrangement of the chromosomes in meiosis. Thus, a particular child could inherit considerably more from the paternal grandfather than from the paternal grandmother.

It is theoretically possible that a child could get only the chromosomes from the paternal grandfather through the sperm of his father, but the chance is so slight as to be considered practically impossible. The chance of any one chromosome being paternal in origin is one half of one and we can find the chance that they would all be paternal by multiplying one half by itself 23 times. The figure obtained would be even smaller if we made allowance for the fact that there can be exchanges of parts of chromosomes during synapsis. The crossing over of chromosomes is discussed in Chapter 11.

MEIOSIS AND INTERSPECIES HYBRIDS

Sometimes closely related animals and plants of different species may be crossed thus yielding a hybrid which is usually sterile. The sterility may be due to an abnormal synapsis of chromosomes in meiosis. For example, the common red fox has 34 chromosomes, whereas the arctic fox has 52. If these two species are crossed, the resulting hybrid will have a diploid chromosome number of 43. When the time comes for meiosis in the reproductive cells of these hybrids, however, normal synapsis is impossible and the distribution of chromosomes to the gametes is so abnormal that the gametes will not function.

This is true even though the chromosome number may be the same in the different species. For instance, both *Drosophila mel-*

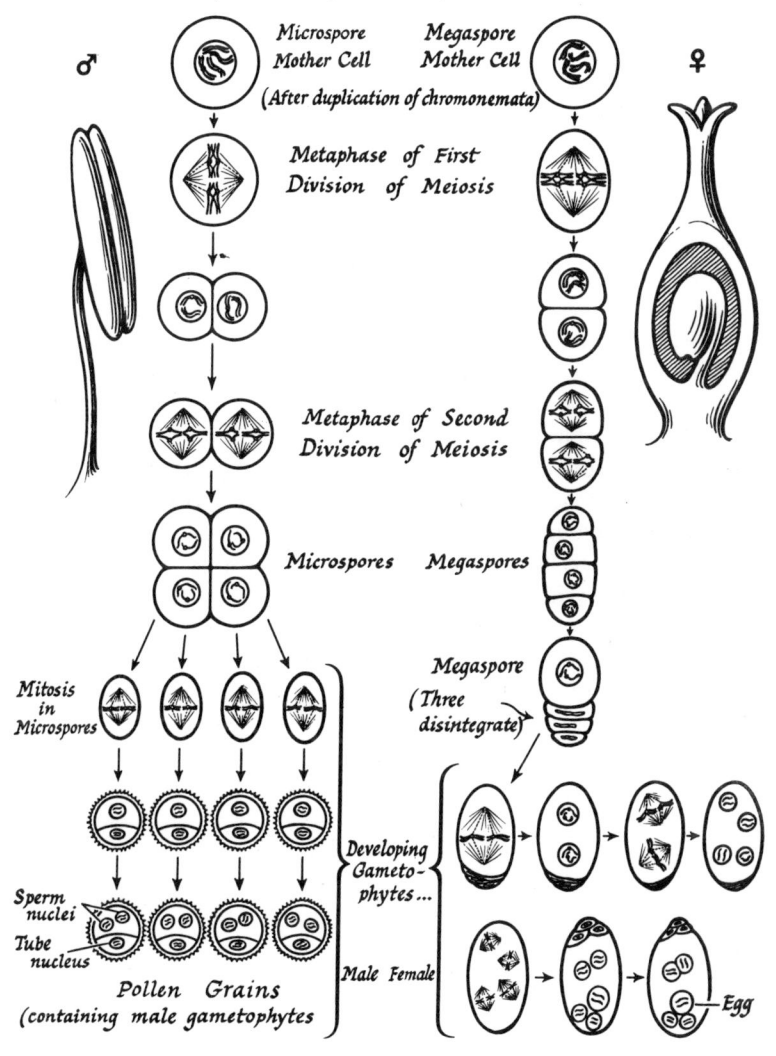

Fig. 3-11. Meiosis in the higher plants. The formation of the male gamete (Microsporogenesis) is shown on the left and the formation of the egg (Megasporogenesis) is shown on the right. (From Winchester, Genetics, Houghton Mifflin.)

anogaster and *Drosophila simulans* have 8 chromosomes in the somatic cells, but the gene arrangement on these chromosomes is so different that normal synapsis in meiosis does not occur and the hybrid is sterile.

MEIOSIS IN PLANTS

In plants the process of meiosis is essentially the same as is found in animals, but there can be a difference in the time in the life cycle when it takes place. It varies according to the classification groups of the plants. In the simpler plants, such as liverworts, meiosis takes place shortly after fertilization and the main body of the plant develops from haploid tissue. Gametes are produced by ordinary mitosis since the cells which produce them are already haploid.

In more advanced plants, such as mosses and ferns, there is a considerable amount of both haploid and diploid tissue. Thus, meiosis takes place at some time between the process of fertilization and gamete production.

In the highest forms of plants, the seed plants, the great bulk of the tissue is diploid as in animals, but there is some nuclear division after meiosis. The process is shown in Fig. 3.11. Since most plants studied genetically are seed plants we can use the same methods to study plant heredity that were used to study animal heredity. As we shall see in the next chapter, the original discoveries concerning the inheritance of unit characteristics were made with plants.

4: THE MONOHYBRID GENETIC CROSS

Since there are thousands of genes on the chromosomes of most organisms, an attempt to study the effects of all of these genes at the same time would be a hopeless task. Nevertheless, it is possible to learn the effects of one gene by carefully controlled breeding of individuals which differ with respect to the gene being studied. In mice there is a characteristic known as kinky tail. We can learn how this is inherited by breeding mice possessing kinky tails with mice having normal tails and observing the offspring for several generations. This procedure is known as a monohybrid cross. So far as possible, we select mice that are alike in all other characteristics and differ only in this one trait. It was through an application of this technique that Gregor Mendel (see Chapter 2) discovered the basic principles of inheritance.

MENDEL'S MONOHYBRID CROSSES

Through numerous monohybrid crosses, Gregor Mendel was able to draw significant conclusions with regard to dominance and recessiveness, the value of large numbers in genetic ratios, gene symbols, and generation symbols.

First Experiments. Mendel selected garden peas for his experiments because they included many pure breeding varieties and, since garden peas normally self-pollinate, their stocks were pure. One of the first of these experiments was concerned with the characteristic position of flowers on the stem. Some plants bear flowers bunched together on the terminal portion of the stem, while others have them along the stem in the axils of the leaves. When Mendel took pollen from a pure-breeding terminal-flowered plant and placed it on the pistil of an axial-flowered plant, he obtained seeds which grew into plants which were axial in every case. But when he allowed these to self-pollinate (the process known as an *inter se cross*) and planted the seeds, he obtained 650 plants with

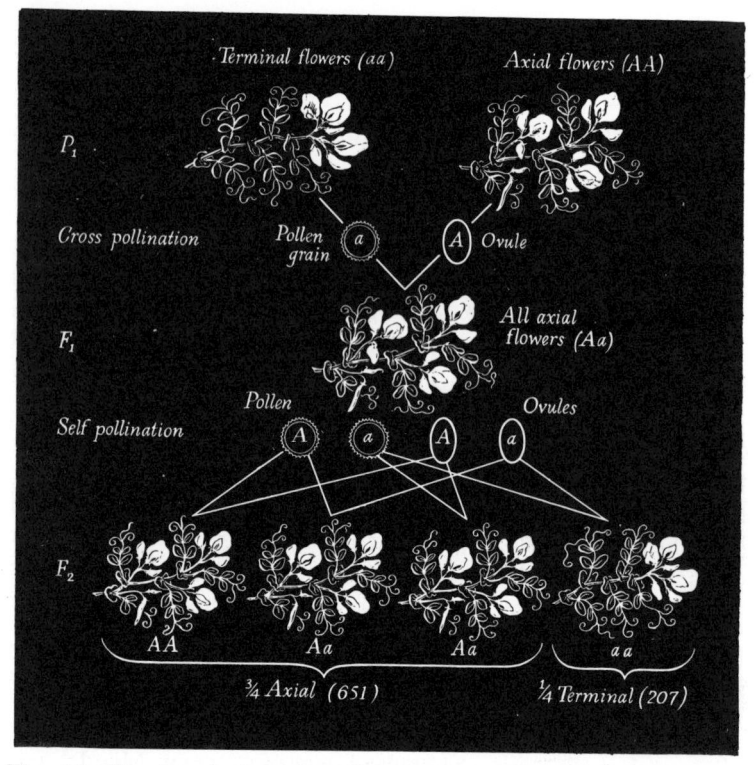

Fig. 4.1. *Results obtained by Mendel when he crossed garden peas having terminal flowers with those bearing axial flowers. It was through crosses of this kind that he worked out the principle of dominance and recessiveness. (From Winchester, Genetics, Houghton Mifflin.)*

axial flowers and 207 with terminal flowers. This is about three fourths axial and one fourth terminal. Mendel made other crosses involving other characteristics and obtained similar ratios. Some of his actual figures are shown in the table on page 43.

Dominance and Recessiveness. After a careful study of these results, Mendel concluded that there must be two "factors" which influence the position of flowers in each plant. One of these, the factor for axial flowers, was dominant; the other, the factor for terminal flowers, was recessive. A plant carrying two factors for terminal flowers would have terminal flowers, but a plant carrying one of the factors for axial as well as one for terminal flowers would bear axial flowers because the factor for axial is dominant

MENDEL'S MONOHYBRID CROSSES

Characteristic	Second Generation Results		Ratio
Form of seed	5474 round	1850 wrinkled	2.96:1
Color of albumen	6022 yellow	2001 green	3.01:1
Color of seed-coats	705 gray-brown	224 white	3.15:1
Form of pods	882 inflated	299 constricted	2.95:1
Color of pods	428 green	152 yellow	2.82:1
Position of flowers	651 axial	207 terminal	3.14:1
Length of stem	787 long	277 short	2.84:1
All combined	14,889 dominant	5010 recessive	2:98:1

over that for terminal flowers. When two of this latter type are crossed, however, there will be a segregation of the factors so that one fourth of the flowers will be terminal flowers, having two recessive factors. Today we use the word "gene" instead of "factor" for the units of inheritance, but Mendel's reasoning was essentially correct.

Value of Large Numbers in Genetic Ratios. We say that the expected ratio in the second generation of Mendel's crosses was 3:1, but a look at the ratios he obtained shows that there was not an exact 3:1 ratio in any cross. A mathematical ratio simply indi-

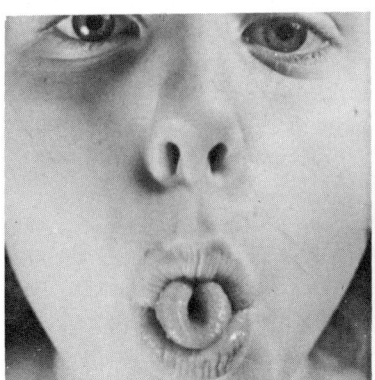

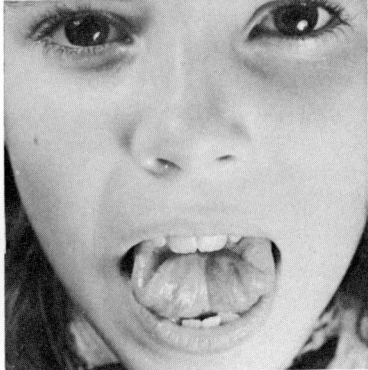

Fig. 4.2. Effects of two dominant genes on the tongue in man. The majority of human beings have the dominant gene which permits them to roll the tongue as shown at left, but less than one person in a thousand has the dominant gene which permits the folding of the tongue as shown on the right.

cates the chance of any one offspring's being one way or the other. Any one plant sprouting from a seed had a ¾ chance of being axial and a ¼ chance of being terminal. Had Mendel planted only a small number of peas, let us say 8, it is entirely possible that he might have obtained 4 axial and 4 terminal plants, a ratio of 1:1. As numbers increase, however, the chances of such wide deviations from the expected ratio decrease. Hence, with 857 offspring, he obtained an actual ratio of 3.14:1 in the flower position crosses; this result is a very close approximation of the calculated ratio. It is also interesting to note that the greatest deviations are to be found in those studies which include the smallest number of offspring. When all are added together, the ratio is 2.98:1, which is about as close to 3:1 as such a count could be. Mendel's awareness of the importance of large numbers probably explains why he succeeded while others failed in attempts to learn the secrets of inheritance through experimental breeding.

Gene Symbols. The system of using letters as symbols for genes, which was devised by Mendel, is employed almost universally today. A small letter stands for a recessive gene, a capital letter of the same kind for the gene which is dominant over this recessive. It is customary to use the first letter of the word which describes the inherited condition that deviates from normal or that is less common. For example, we use O as the symbol for the dominant gene which causes *otosclerosis*, a type of inherited deafness in man, and *o* to represent the recessive gene which functions in producing normal hearing. Likewise we use *a* for the recessive gene for albinism (defective deposition of pigment in the skin, hair, and eyes) and A for its counterpart which functions in normal pigment production.

When we are working with animals in which we have discovered a large number of genes and wish to use the same letters for the same genes consistently, we may extend the gene symbols to two or three letters. With *Drosophila*, for example, we use the symbol *vg* for the gene which produces vestigial wings and *car* for the gene which produces carnation (color) eyes. Various other modifications will be introduced as we continue our study.

Generation Symbols. Mendel also introduced the various symbols which are in use today as a means of designating the generations. He designated the plants with which he started a cross as the P_1, or first parental generation. The offspring from these parents were designated as the F_1, or first filial generation, and the offspring of these were designated as the F_2 second filial generation. If the breeding was carried further it is possible to use F_3 and so on.

MECHANISMS OF THE MONOHYBRID CROSS

Mendel's principles as worked out on garden peas have been found to be generally true for most other forms of life. An illustration with rabbits will show how they are applied to animals. We can select a male rabbit with a black coat that has descended from a pure strain in which all the members had black coats for many generations. This indicates that he carries two genes for black— we say that he is *homozygous* black. We breed this male with a homozygous brown female. All the sperms carry a gene for black, and all the eggs carry a gene for brown. Thus, all offspring will have one gene of each kind. They will all be black because the gene for black is dominant and the gene for brown is recessive. Since they carry the gene for brown, however, these are called *heterozygous* black rabbits.

The F_1 rabbits produce two kinds of gametes. One half of the gametes will carry a gene for black and the other half will carry a gene for brown. Crossing two of these gametes results in the following gene combinations: (1) The sperm which has the gene for black may unite with the egg which has the gene for black and yield a homozygous black rabbit. (2) The sperm which has the gene for black may unite with the egg which has the gene for brown and yield a heterozygous black rabbit. (3) The sperm which has the gene for brown may unite with the egg which has the gene for black and thus produce a second heterozygous black rabbit. (4) Finally, the sperm which has the gene for brown may unite with the egg which has the gene for brown and yield a brown rabbit. This gives a ratio of three black rabbits to one brown rabbit. Two of the black rabbits are heterozygous and one is homozygous, while the brown rabbit is homozygous, as must always be the case where a recessive gene is expressed.

Genotype and Phenotype. These are terms which are convenient to use in genetic discussions. Genotype simply means the type of genes present in the organism. It is generally expressed by means of the letter symbols of the genes carried by the organism. Pheno-

46 THE MONOHYBRID GENETIC CROSS

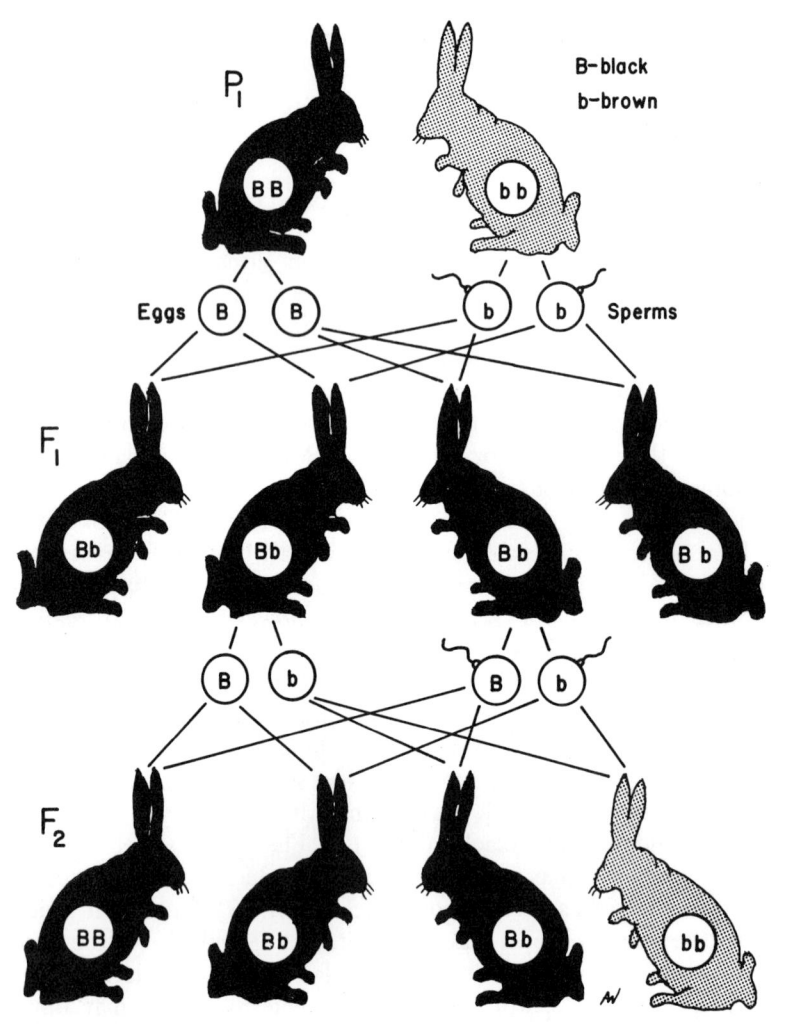

Fig. 4.3. A cross between black and brown rabbits illustrating the mechanism of the monohybrid cross.

type refers to the expression of the genes. For example, the phenotype of the F₁ rabbits is black—the genotype is *Bb* (*B* symbol being the dominant expressed trait, and *b* the recessive trait), or heterozygous black.

Alleles. Each gene has a particular position on a particular chromosome at its **locus** (location). In some animals which have been extensively studied by geneticists, the locus can be identified through special breeding and cytological techniques. In *Drosophila melanogaster*, for example, there is a gene which functions in wing production that is known to be located about one third of the distance from one end of a chromosome known as chromosome II. Since there are two chromosomes of each kind in a somatic cell, there will be another gene affecting wing growth at the same locus on the other chromosome II. Two varieties of this gene are known to exist; one functions normally, contributing its part in producing a normal long wing. The other, however, exerts an effect on wing development which prevents normal wings from forming, and the flies thus affected have mere stumps of wings, vestigial wings. This second gene, with the symbol *vg*, is recessive to the dominant gene for normal wings, with the symbol *Vg*.

We say that these two genes are alleles because they occupy the same position on the chromosome and will come together when the two chromosomes enter into synapsis during the first division of meiosis. There are other genes which affect wing growth at other positions on the chromosomes—one of these is a

Fig. 4.4. Fruit fly with vestigial wings. This condition is brought about by a recessive gene which is an allele of a gene on chromosome II, which functions in the production of normal wings.

recessive gene for dumpy wings (*dp*) which produces wings about half the normal size. This gene, however, is located near the opposite end of chromosome II. Hence, it would not be an allele of the vestigial-producing gene. One way to demonstrate this is to cross a homozygous vestigial-winged fly with a homozygous dumpy-winged fly. All of the F_1 are normal because each fly carries the dominant normal alleles of the other wing characteristic. They would have the genotype *Dp vg/dp Vg*.

It might be helpful at this point to mention the fact that frequently in *Drosophila* breeding the plus sign is used for the normal allele. This is the gene which would be found in the great majority of wild flies captured away from the laboratory. Thus we might write the heterozygous genotype as $+ dp/vg +$.

Diagramming Monohybrid Crosses. Fig. 4.3 shows the method of working out crosses through the use of letters for genes. This shows the diploid parents, the haploid gametes, and the combinations of the gametes to yield the diploid offspring. As you gain a better understanding of the process, you may wish to simplify such a diagram and represent it as follows:

$$Bb \times Bb = BB + 2Bb + bb \text{ (3 black : 1 brown)}$$

This gives the minimum essentials—the genotype of parents and offspring and the phenotype of offspring.

A cross between a heterozygous black rabbit and a brown rabbit would be:

$$Bb \times bb = Bb + bb \text{ (1 black : 1 brown)}$$

Actually, we would obtain two of each of these genotypes in the offspring if we showed all possible combinations of genes, but it is simpler to write this ratio as 1:1 rather than 2:2.

THE TEST CROSS

In the examples used to illustrate the mechanism of the monohybrid cross, we have assumed that we knew the genotype of the parents and that we have determined the phenotype of the offspring through proper combinations of the parent reproductive cells. In actual practice, however, it frequently happens that we can observe the phenotype with no trouble, but we are not sure of the genotype of the individuals expressing a dominant gene.

For example, a cocker spaniel dog breeder may want his dog to produce only pups with a solid body color, but he finds that sometimes he gets some pups with white areas on the body (parti-colored). We know that this trait is due to a recessive gene and it is evident that the man has some heterozygous dogs in his breeding stock. He never breeds any parti-colored dogs, but this is not

THE TEST CROSS

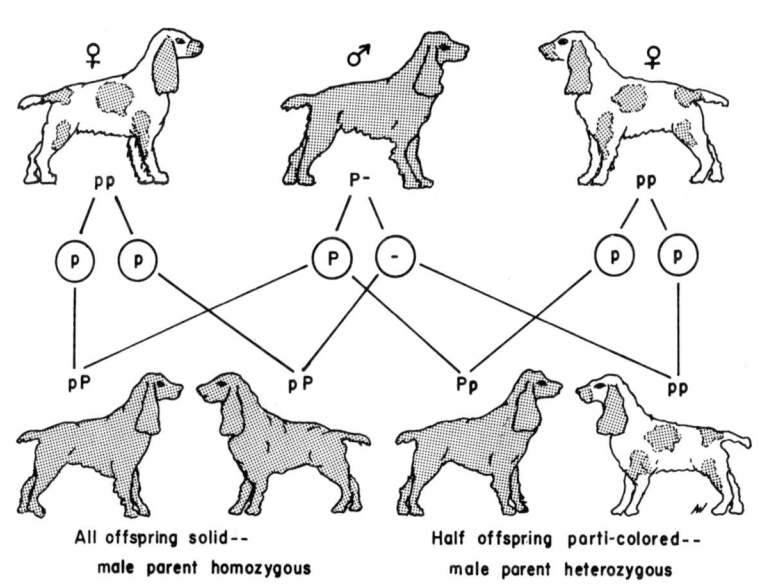

Fig. 4.5. *The test cross illustrated with a coat characteristic in cocker spaniels. The test is being made on the male in the center to determine if he carries the recessive gene for parti-colored body coat. He is crossed with a parti-colored female. If he does not carry the recessive gene all his offspring will have a solid coat as shown by the cross on the left. If he carries the gene for parti-colored coat about one half of his offspring will be parti-colored as shown by the cross on the right. A dash represents the unknown gene of the male.*

sufficient to eliminate the gene as it continues to be carried by heterozygous dogs. He can use a test cross, however, to isolate and eliminate the heterozygous dogs.

To do this he would select a male with a solid body color. This dog would have the genotype P/–. The dash indicates that the second gene is unknown. This male is bred with a parti-colored female which we know would have the genotype p/p. The phenotype of the offspring resulting from this cross will show the nature of the unknown gene in the male. If the offspring are about one half parti-colored, then it is evident that the male is heterozygous and should not be used for further breeding. On the other hand, if all the pups are solid colored it would appear that the male is homozygous and could serve as the father of a pure breeding stock. Females could likewise be tested and only homozygous females selected for breeding. Thus, the parti-colored gene could be eliminated from his breeding stock. These crosses are shown in Fig. 4.5.

GENES WITH INTERMEDIATE EXPRESSION

The specific factors included under genes with intermediate expression are absence of dominance, determination of genotype, and incomplete dominance.

Absence of Dominance. Sometimes when two different alleles come together there is an absence of dominance and both genes are partially expressed, giving an effect intermediate between the phenotypes which appear when either of the genes are homozygous. In short-horned cattle there are allelic genes for white and for red hair color. If we cross a red bull with a white cow, the offspring are roan in color—this is intermediate between the two parents. Selecting letter symbols for such intermediate genes is somewhat difficult. The use of small and large letters would seem to indicate that one gene is dominant and the other recessive. We can solve the dilemma by using an I to indicate intermediate and a superscript to indicate the characteristic. This diagram of a cross between two roan cattle will illustrate:

$$I^r I^w \times I^r I^w = I^r I^r + 2\, I^r I^w + I^w I^w \ (1 \text{ red} : 2 \text{ roan} : 1 \text{ white})$$

Determination of Genotype. We have no problem in recognizing the genotype in cases of intermediate expression. The heterozygous individuals stand out clearly from either homozygous phenotype. In the domestic fowl (chickens) there is a breed known as the Andalusian which is a blue color. Whenever two of these are bred, the offspring are black, blue, and a splashed white in a ratio of 1:2:1. It is clearly evident that the blue color must

INTERMEDIATE INHERITANCE

Fig. 4.6. *Intermediate expression of genes in chickens. The heterozygous individual produced in this cross of Andalusian chickens has a steel-blue color.*

be a heterozygous expression of the genes for black and splashed white (white with splashes of black).

Manx cats have short tails and there is a considerable demand for this type of cat. The short tail, however, results from an intermediate expression of alleles concerned with tail growth. In breeding two Manx cats only one half of the offspring will be Manx—the others with long tails and those with no tails will not find a ready market. The breeder can obtain all Manx kittens, however, if he uses only long-tailed toms and no-tailed females, or vice versa.

Incomplete Dominance. Dominance is a relative thing—in some cases it appears to be complete with no distinguishable difference between the homozygous and the heterozygous individuals. In other cases, as we have shown, the effect is approximately between the two allelic genes, thus producing an intermediate effect. There are some cases noted, however, wherein one gene has a much more pronounced effect than the other, but the heterozygous individual expresses a detectible phenotypic effect of the other gene. A careful search for such small heterozygous expressions reveals many cases which come under this category. In fact, some geneticists feel that if we can devise sufficiently sensitive tests we may find that the majority of allelic genes are included.

As an example, there is a recessive gene in man for *xanthomatosis*. When an individual is homozygous, this gene causes a defect in fat metabolism and there is an excessive accumulation of lipoids in various body parts. These tend to form fatty tumors which reduce the general body health of the affected person. Recently, it has been discovered that heterozygous persons have an unusually high cholesterol content of the blood. Thus, it is possible to recognize carriers of this recessive gene and predict the appearance of xanthomatosis in the children.

Sometimes the heterozygous expression seems to bear little relationship to the homozygous recessive phenotype. *Friedreich's ataxia* in man is an affliction characterized by a failure of muscular co-ordination and an impairment of speech. The disorder appears when a person is homozygous for a certain recessive gene. Whenever a child has this condition it is found that the normal parents, who must be heterozygous, have an excessive curvature of the sole of the foot (*Pes cavus*). This apparently unrelated characteristic is an indication of the presence of one of the recessive genes. It is easy to see how knowledge of this kind can be of great value in human genetics and in plant and animal breeding.

GENES AFFECTING VIABILITY

As different kinds of genes are studied it becomes evident that there are some genes which tend to reduce the chance of survival of the organisms which express them. The effect may vary from a slight lowering of viability to an effect so great that all individuals that express the gene will die.

Lethal Genes. These genes may exert their effect early in life so that death comes when the embryo is only a microscopic bit of protoplasm. In other cases the embryo may develop normally for a time and death comes at a later stage of embryonic development. In man, for example, a gene which prevents normal kidney function would not cause the death of the embryo since the mother's kidneys carry out the process of waste removal from the fetal blood stream. At birth, however, the improper functioning of the kidneys will result in death. Albinism in plants does not interfere with embryonic growth and the sprouting of the seed, but when the food which was present in the seed is exhausted the plant will die, because in the absence of chlorophyll it cannot manufacture its own food.

Frequency of Lethal Genes. From mathematical calculations it is estimated that, on the average, each person will carry about four lethal genes. However, these genes do not exert their effect because there are normal alleles which prevent their full expression. Moreover, there are thousands of possible lethal genes, and at marriage it is very unlikely that both marriage partners will carry the same lethals. When they do, the lethal gene they hold in common will be expressed in about one fourth of their children. When the genes are expressed in this way they are eliminated from the race, but the number of lethals remains about constant because the rate of mutation produces new lethals at the same rate as they are eliminated in a stable population. It is estimated that about one fourth of all mutations are lethal.

In marriages between relatives, such as marriages between cousins, the chance that both marriage partners will carry the same lethal gene will be greatly increased because of their common ancestry. Hence, in any group where there is widespread inbreeding there will be a more rapid elimination of lethals. Also, any factor which changes the rate of mutation, such as widespread radiation exposure, will increase the pool of lethals in the population.

Intermediate Lethals. It is obvious that dominant lethal genes cannot exist in any germ plasm. Any dominant lethals which appear will be automatically eliminated in the individuals in which they arise. We have learned, however, that some genes have an effect when heterozygous which is not as extreme as when they are homozygous. Likewise, there are some lethals with somewhat of an intermediate effect that shows phenotypically in the heterozygote, but whose effect is not lethal. A classical illustration of this condition is found in cattle. In some cattle in England there is a lethal gene which, when homozygous, produces what is known as a bulldog calf. Such a calf has very short legs, a shortened muzzle which gives it the appearance of a bulldog, and other deformities which cause death at, or shortly after, birth. When the gene is heterozygous, cattle are produced with short legs and other smaller expressions of the gene, but they are perfectly viable and have good beef-producing qualities. These are known as **Dexter cattle. Kerry cattle** are homozygous for the allele of this lethal and they have normal legs and other characteristics. When two Dexters are crossed they yield offspring in the ratio: 1 Kerry : 2 Dexter : 1 Bulldog. Since the bulldog calf dies the ratio of the surviving offspring is 1 Kerry : 2 Dexter.

In many cases, of course, it is not possible to recognize the homozygous lethal because of early embryonic death and its presence can only be ascertained by (1) the unusual ratios of living offspring which are obtained and by (2) a reduction of the total number of expected offspring. In chickens, for example, about one fourth of the eggs would not hatch; in dogs the size of the litters would be reduced by about one fourth; in corn about one fourth of the seeds would not sprout. A typical example of this type of intermediate lethal is the gene for **Dichaete wings** in *Drosophila*. This gene, when heterozygous, causes the wings to be held out to the side instead of back over the body and also causes the bristles on the body to be shorter than normal. It is lethal when homozygous, but the embryo does not survive long enough to be detected. Hence, a cross of two flies with a Dichaete wing phenotype will yield one third normal flies and two thirds Dichaete. Dichaete crossed with wild type yields one half of each. These ratios are an indication of an intermediate type lethal.

Intermediate Lethals in Man. There are probably many intermediate lethals in man. Many of the rare genes for abnormalities which we now class as dominants might be of this nature. The lethal effect of the homozygous condition has not been observed because the genes are so uncommon that we have never found families where both parents show the heterozygous phenotype. One pedigree which has been found can serve to illustrate this. The gene for **brachyphalangy** causes a shortening of the fingers. Heterozygous persons appear to have only two joints to the fingers because the middle joint is greatly shortened and often fused to one of the other finger bones. It was thought to be a simple dominant gene until one geneticist found a marriage where both parents had this condition. Of their four children one was born without any fingers or toes and with other skeletal defects and was

GENES AFFECTING VIABILITY

Intermediate Killer Genes...
When two people with Brachyphalangy marry

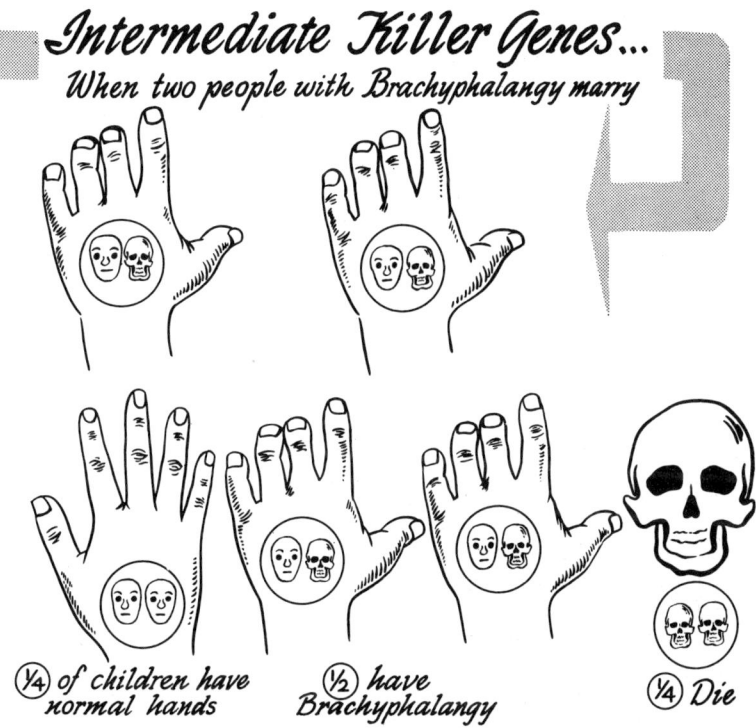

(¼) of children have normal hands   *(½) have Brachyphalangy*   *(¼) Die*

Fig. 4.7. Effect of an intermediate type of lethal gene in man. A marriage between two persons with brachyphalangy resulted in a distribution of children as shown in this diagram. (From Winchester, Heredity and Your Life, Dover Press.)

unable to survive. Two other children possessed the short fingers; one child was normal. This is the exact 1:2:1 ratio that would be expected if we consider the short fingers to be an intermediate expression of a lethal gene.

TYPICAL PROBLEMS AND ANSWERS

1. In domestic poultry some chickens have a rose comb (multiple divisions) and some have a single comb. Explain how you could determine which of these characteristics is dominant and which is recessive.

Answer. Cross a rooster from pure-breeding rose-combed stock with hens from pure-breeding single-combed stock. Then cross the F_1

among themselves and tabulate the F_2. If all of the F_1 and ¾ of the F_2 are rose and ¼ single, then we would conclude that a rose comb is caused by a simple dominant gene and a single comb by a recessive gene. The reverse results would indicate that single is dominant and rose recessive.

2. Rose comb is the dominant characteristic. Cross a homozygous rose-combed rooster with a homozygous single-combed hen and show the offspring. Also, show the offspring of a cross of two of the F_1.

Answer.

F_1 $RR \times rr = Rr$ (all rose-combed)

F_2 $Rr \times Rr = RR + 2Rr + rr$ (3 rose : 1 single)

3. Consider blue eyes in man as recessive to brown eyes. Show the expected children of a marriage between a blue-eyed woman and a brown-eyed man who had a blue-eyed mother.

Answer. The man evidently would be heterozygous since he would have received a gene for blue from his mother. Hence the cross would be:

$bb \times Bb = bB + bb$ (1 blue : 1 brown)

4. In domestic swine there is a dominant gene which produces a white belt around the body, while the recessive allele results in a uniformly colored body. One farmer wants to produce only belted hogs; another wants only solid colored. Which farmer would have the easier task of establishing a pure-breeding stock? Tell how each would proceed.

Answer. It would be easier to produce the pure-breeding solid colored hogs, for it is only necessary to select solid colored hogs for parents. To get pure-breeding belted hogs, on the other hand, the farmer must test his males for homozygosity by breeding them to solid colored females. The males that produce any solid colored offspring would be discarded as they would be heterozygous. The same test would be applied to the females.

5. A man crosses two pink-flowered four-o'clocks and obtains seeds which grow into plants bearing the following flower colors: 24 red, 53 pink, and 28 white. From these results tell how you think these colors are inherited and show a diagram of the cross.

Answer. Since these results give an approximate 1:2:1 ratio, it appears that this is a case of intermediate expression, absence of dominance, with pink as the heterozygous expression of genes for red and white.

$$I^r I^w \times I^r I^w = I^r I^w + 2 I^r I^w + I^w I^w \text{ (1 red : 2 pink : 1 white)}$$

6. In cattle the polled (hornless) condition is due to a dominant gene, while its recessive allele causes horns to appear. Two polled cattle have a calf which develops horns as it matures. Show the genotypes of all three.

Answer.   Parents: Pp and Pp Calf: pp

7. In man there is a gene that, when homozygous, causes a severe anemia known as sickle cell anemia, so named because many of the red cells are abnormal and assume a sickle shape. Death nearly always occurs before adulthood. Heterozygous persons appear normal, but when the blood is held under a low oxygen concentration the red cells become sickle-shaped also. A young woman, about to be married, has a brother who died of sickle cell anemia and she is concerned about the chance of the condition appearing in her children. When blood samples are taken and placed under low oxygen concentration, her blood becomes sickled, but that of her prospective husband remains normal. What could you tell her about her children?

Answer. The fact that her blood became sickled shows that she carries the gene (is heterozygous), but since her fiancé's blood remains normal, he would be homozygous for the normal gene. Thus, no children would have the anemia, but about half would carry the gene.

8. To obtain Dexter cattle, breeders sometimes cross two Dexters. This cross gives one half Dexter offspring, but one fourth of the calves are bulldog and die. Explain how Dexters could be obtained without any loss of calves.

Answer. Cross Dexter with Kerry cattle. This will give one half Dexters, but the other half are Kerrys and there are no bulldog calves. You obtain just as many Dexters from this cross and there is no loss of calves. The cross is diagrammed as follows:

I^K Kerry I^B Bulldog $I^K I^B$ Dexter

$$I^K I^K \times I^K I^B = I^K I^K + I^K I^B \text{ (1 Kerry : 1 Dexter)}$$

9. Albinism in corn is due to a recessive gene which is normally lethal because the plant cannot manufacture food without the green

chlorophyll and dies as soon as it exhausts the food stored in the seed. It is possible, however, to keep albino plants alive by special feeding techniques by means of which sugar is supplied to the plant through the leaves. Show the expected offspring from a cross between such an albino plant and a normal plant heterozygous for the gene for albinism.

Answer.

$$aa \times Aa = aA + aa \text{ (1 normal : 1 albino)}$$

5: THE DIHYBRID GENETIC CROSS

We have gained some insight into the method of behavior of one pair of genes during gamete formation and the production of zygotes through fertilization. Now we will learn how these genes react with respect to other genes in the same individuals. Mendel realized the importance of such information and made dihybrid crosses (crosses involving two pairs of genes) in an effort to obtain it.

MENDEL'S DIHYBRID CROSSES

Mendel's experiments with peas laid the basis for his formation of a new genetic principle.

Yellow-round and Green-wrinkled Peas. One of Mendel's first dihybrid crosses involved the shape and color of the seeds drying. Some remained round when dry while others shriveled and were wrinkled. Some were yellow and some were green. When homozygous yellow-round peas were crossed with homozygous green-wrinkled peas, the F_1 were all yellow-round. This showed that the gene for yellow and the gene for round are dominant. When these were allowed to self-pollinate, the F_2 yielded 315 yellow-round, 101 yellow-wrinkled, 108 green-round, and 32 green-wrinkled.

An analysis of these figures reveals an approximate 9:3:3:1 ratio. Thus, we can see that Mendel obtained not only the parental types, but also two new combinations of characteristics which resulted from mixing of the characteristics found in the P_1.

A dihybrid is nothing more than two monohybrids acting independently. If we consider the F_2 peas in terms of the color of the seed, the numbers obtained were 416:140, while for round or wrinkled the numbers were 423:133. In both cases we have a close approximation of the 3:1 ratio.

Principle of Independent Segregation. These results led Mendel to formulate his law of independent segregation which holds that (1) there is free assortment of the genes in gamete formation and (2) new combinations of characteristics can be found in the offspring.

THE DIHYBRID CROSS DIAGRAM

It is necessary to consider here the principal methods of diagramming dihybrid crosses, including the method of Punnett's Squares.

The Checkerboard Method (Punnett's Squares). In diagramming dihybrid crosses we run into some difficulty of mechanics which we did not have in the monohybrid. If we cross two homozygous individuals the gametes of each are all alike and the F_1 are all of the same genotype—this presents no problem. As we go into meiosis and separation of the genes of these F_1 individuals, however, we find that there are four possible combinations of genes. There are two combinations the same as were found in the parents and, there are also two new combinations. We can handle these by drawing a "checkerboard" with four squares in each direction and placing the male gametes along one side and the female gametes along another side, as shown in Fig. 5.1. This procedure was first worked out by Punnett in England; hence the diagrams are sometimes called Punnett's squares. By making the proper combinations we can obtain the genotype of the offspring and the proportion of each. Since we find that 9 squares have at least one dominant gene of each type, we expect $9/16$ yellow-round. Since one square bears two recessive genes of both types, we expect only $1/16$ to show both recessives. The other two combinations of $3/16$ each are obvious.

Other Types of Dihybrid Crosses. In working out dihybrid diagrams, it is not always necessary to show the complete checkerboard of 16 squares. For example, if we have a cross of a guinea pig heterozygous for the dominant gene for black hair, but homozygous for the dominant gene for short hair, $W/w \ L/L$, there are only two types of gametes possible: WL and wL. If we crossed such a guinea pig with one heterozygous for both characteristics, we would need to show only two rows of four squares each in order to determine the genotype of the offspring.

Should we desire to cross a heterozygous black, long-haired guinea pig with a white, short-haired guinea pig (homozygous, of course), we would need only one row of four squares or we could

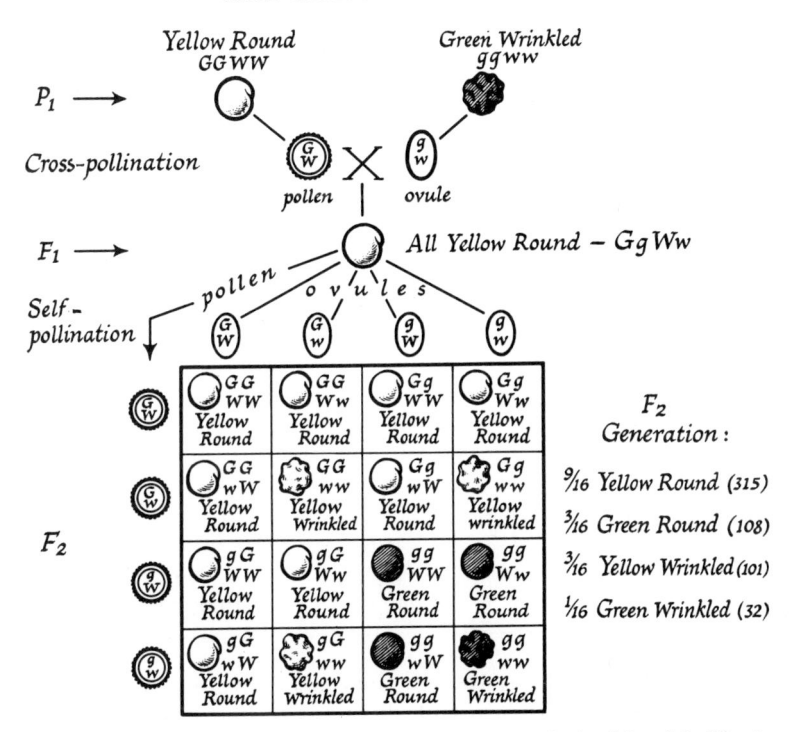

Fig. 5.1. A dihybrid cross in garden peas as made by Mendel. Plants bearing yellow-round seeds were crossed with those bearing green-wrinkled seeds. (From Winchester, Genetics, Houghton Mifflin.)

go back to our monohybrid method of drawing circles and connecting lines. Such a cross (the test cross) is used very often in dihybrid genetic studies because it gives a ratio of 1:1:1:1 rather than the 9:3:3:1 which would be expected from the *inter se* type of cross.

MORE COMPLEX HYBRID CROSSES

It is possible to make crosses of individuals differing with respect to more than two characteristics. A trihybrid would involve three characteristics with eight possible types of gametes for each heterozygous individual. This requires a checkerboard with eight rows of eight squares, a total of 64 squares, and yields a phenotypic ratio of 27:9:9:9:3:3:3:1. With four characteristics there would be 16 types of gametes, and so on. It is obvious that such figures would soon become unmanageable. The following table shows how rapidly the numbers increase.

Pairs of Hetero-zygous Genes	Types of Gametes	Possible Zygotic Combinations
1	2	4
2	4	16
3	8	64
4	16	256
5	32	1,024
6	64	4,096
7	128	16,384
8	256	65,536
9	512	262,144
10	1,024	1,048,576
15	32,768	1,073,741,824
20	1,048,576	1,099,511,627,776

This table well illustrates how the genetic diversity of organisms can be so great that, even with so small a number of genes as 20 pairs, the chance of two offspring receiving the same gene combinations from their parents is considered to be so improbably small as to be virtually impossible.

MODIFIED DIHYBRID RATIOS

Two major factions may produce variations in the dihybrid ratios: intermediate genes, and inhibiting genes (epistasis).

Effect of Intermediate Genes. Just as the monohybrid cross of heterozygous individuals may yield offspring which vary from the 3:1 ratio, so may the dihybrid vary from the 9:3:3:1 ratio. Intermediate genes can be one cause of such variation. In cattle the polled (hornless) condition is due to a dominant gene, while horns result from a pair of recessive genes. The red and white coat colors are intermediate and give a roan color when the genes are heterozygous. Thus we have three possible coat colors and two conditions affecting the horns. A cross between homozygous polled-red cattle and horned-white cattle will produce all polled-roan offspring. An *inter se* cross will give an F₂ with the ratio of 6:3:3:2:1:1. Thus, there are six classes of offspring rather than four.

An even greater number of classes appears if both genes are intermediate. This can be illustrated by the snapdragon if we use the color of the flowers and the width of the leaves as characteristics. When we cross a red-flowered, broad-leaved plant with a white-flowered, narrow-leaved plant, we obtain all pink-flowered, intermediate-leaved plants in the F₁. In the F₂ we find that we have just as many phenotypes as we have genotypes—the ratio is 4:2:2:2:2:1:1:1:1. Such unusual ratios should give little difficulty if you use the checkerboard and interpret the phenotype of each square correctly.

EPISTASIS

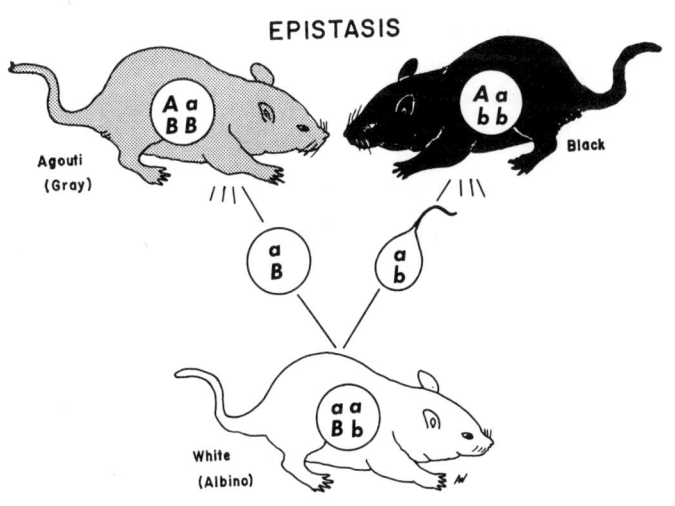

Fig. 5.2. White offspring from a cross of an agouti with a black mouse. Both parents carry the gene for albinism and when the gene becomes homozygous in an offspring the coat is white because the gene for albinism is epistatic to the other genes which determine coat pigmentation.

Epistasis (Inhibiting Genes). Epistasis is another factor which can introduce variation in the dihybrid ratio. We can illustrate with a cross in mice. In a cross that took place between a mouse homozygous for the dominant wild type agouti (gray) coat and a mouse that was homozygous for the recessive black coat, we obtained the surprising F₁ ratio of 3 agouti : 1 white (albino).

Usually a cross between such mice will give all agouti offspring because agouti is dominant over black. In this case, however, both mice were heterozygous for the recessive gene for albinism. When homozygous, this gene prevents the formation of pigment in the hair; therefore, it makes no difference what the genotype for hair color might be—the mouse will be white. The gene for albinism is not an allele of the gene for black or agouti; hence the example given was a dihybrid cross with the following genotypes.

$$B/B\ A/a \times b/b\ A/a = B/b\ A/A + B/b\ A/a + B/b\ A/a + B/b\ a/a$$
$$\ (3\ \text{agouti}) (1\ \text{white})$$

Thus, we can say that the gene for albinism is epistatic to the genes for the different kinds of coat pigmentation. Since it requires two genes, however, for the albinism to be expressed, we call this a case of *recessive epistasis.*

Dominant epistasis is found when only one gene of a pair of alleles is required to mask the influence of a second pair of genes. We can illustrate with some odd results from the domestic fowl. When white Plymouth Rock chickens are crossed with white Leghorn chickens, the F_1 offspring are all white as would be expected, but the F_2 give a ratio of 13 white : 3 colored. This is accounted for by the genotype of the two breeds of chickens. White Leghorns are homozygous for a dominant gene C for color, but they are also homozygous for a dominant I which inhibits color formation. (I is epistatic to C.) Plymouth Rocks, however, are white because they lack the dominant gene for color and are homozygous for the recessive gene for lack of color, c. They are also homozygous for i, which would allow color to form if there were any genes for color. The cross between these two would be: $C/C\ I/I \times c/c\ i/i = C/c\ I/i$. These are all white because they have the inhibitor gene and this gene is dominant. In the F_2, however, there will be 3 with the gene for color that do not also have the gene I which inhibits color formation $(C/-i/i)$. These will be colored.

Duplicate recessive epistasis may be illustrated by deaf-mutism in man. There are two recessive genes, either of which can cause deaf-mutism when homozygous. Hence, two deaf-mutes could marry and have children all of whom are normal provided that the deafness had been caused by the action of two different genes. In marriage between two persons with such parents, however, one

MODIFIED DIHYBRID RATIOS 65

EPISTASIS IN DEAF-MUTISM

A & B necessary for normal hearing
a or b deaf-mutism when homozygous

Heterozygous Normal Parents: A a B b × A a B b

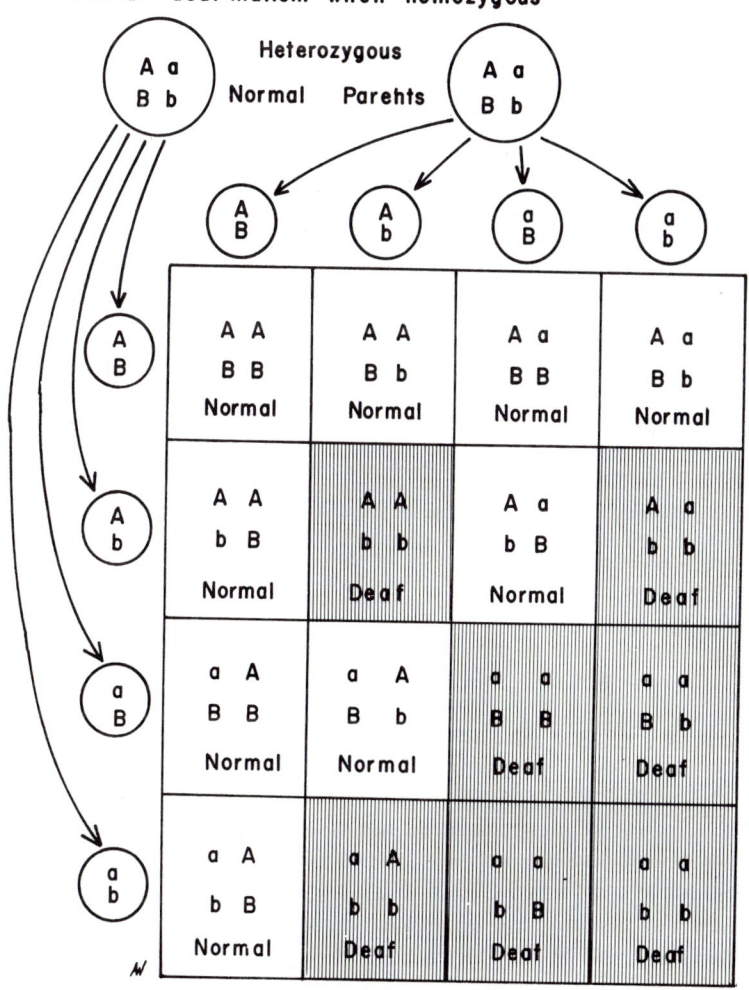

Gametes: AB, Ab, aB, ab

	AB	Ab	aB	ab
AB	AA BB Normal	AA Bb Normal	Aa BB Normal	Aa Bb Normal
Ab	AA bB Normal	AA bb Deaf	Aa bB Normal	Aa bb Deaf
aB	aA BB Normal	aA Bb Normal	aa BB Deaf	aa Bb Deaf
ab	aA bB Normal	aA bb Deaf	aa bB Deaf	aa bb Deaf

Fig. 5.3. *Deaf children from parents with normal hearing.* When both parents are heterozygous for two recessive genes for deaf-mutism they may bear afflicted children as shown above. The unusual ratio of 9 normal : 7 deaf children results because both recessive genes are epistatic to the genes for normal hearing. This is a case of duplicate recessive epistasis.

would expect a 7/16 chance of deaf-mutism in their children. This is illustrated in Fig. 5.3.

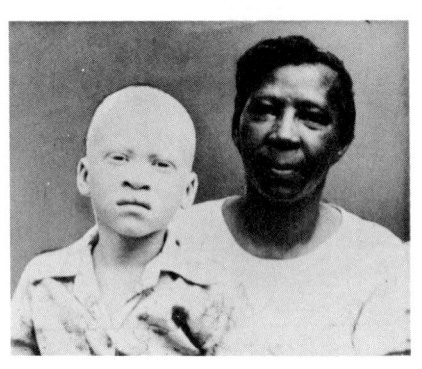

Fig. 5.4. *A white child with Negro parents. This boy shown with his mother is homozygous for the gene which causes albinism and since this gene is epistatic to the genes which produce the normal negroid pigmentation, he has the very fair skin of the albino. Both parents carry the gene in the heterozygous state but, since it is recessive, they have the Negro pigmentation.*

Finally, we will mention *duplicate dominant epistasis,* which requires only one of either of two dominant genes for inhibition of the effect of another gene at a different locus. Almost all chickens have unfeathered shanks and are homozygous for the two recessive genes which are responsible for this characteristic $a/a\ b/b$. Either a dominant A or a dominant B, however, will cause the shanks to be feathered. A cross between two heterozygous chickens yields 15 feathered:1 unfeathered. Only the square of the checkerboard containing all four recessive genes will result in the unfeathered shanks.

SHORT CUTS IN OBTAINING PHENOTYPIC RATIOS

Although every student should have some practice in working out dihybrid ratios through the use of the checkerboard, it is possible to use shorter methods when only the phenotypic ratio is desired. We do this by considering the dihybrid as two monohybrids. For instance, we know that a cross between two heterozygous black guinea pigs yields the phenotypic ratio of 3 black : 1 white. Also, we know that a similar ratio is obtained in a cross between two heterozygous short-haired guinea pigs (3 short : 1 long). We can expect to find the number of black short from a dihybrid cross by multiplying the chance of each in the monohybrid cross—3 black $\times$ 3 short = 9 black short. Similarly, 3 black $\times$ 1 long = 3 black long. The entire ratio can be represented as:

3 black + 1 white
× 3 short + 1 long
─────────────────
9 black short + 3 white short + 3 black long + 1 white long

Or, it is possible to use gene symbols as follows, even though there is little saving in time or space by this method.

3 W– + 1 ww
× 3 L– + 1 ll
─────────────────
9 W– L– + 3 ww L– + 3 W– ll + 1 ww ll

The method applied to a more complex dihybrid involving a cross between roan cattle heterozygous for polled would be as follows:

1 red + 2 roan + 1 white
× 3 polled + 1 horned
─────────────────
3 red polled + 6 roan polled + 3 white polled + 1 red horned
 + 2 roan horned + 1 white horned

We can even use this method in cases of epistasis if we keep in mind which gene is epistatic to the other. For the case of marriage between persons with deaf-mute parents it would be:

3 normal (A–) + 1 deaf-mute (aa)
× 3 normal (B–) + 1 deaf-mute (bb)
─────────────────
9 normal (A–B–) + 3 deaf-mute (aaB–)
 +3 deaf-mute (A–bb) + 1 deaf-mute (aabb)

9 normal : 7 deaf-mutes

The duplicate dominant epistasis as illustrated by the feathered shanks in chickens would be:

3 feathered (A–) + 1 smooth (aa)
× 3 feathered (B–) + 1 smooth (bb)
─────────────────
9 feathered (A–B–) + 3 feathered (aaB–)
 +3 feathered (A–bb) + 1 smooth (aabb)

15 feathered : 1 smooth

To use this method be sure to start with the phenotype which would result from each gene as a monohybrid and then multiply, making due allowance for epistasis if such is present.

TYPICAL PROBLEMS AND ANSWERS

1. In dogs the tendency to bark while trailing is due to a dominant gene while silent trailing is recessive. Erect ears is dominant over drooping ears. Show the offspring expected from a cross between two erect-eared barkers who are heterozygous for both genes $(D/d\ B/b)$.

Answer. Using the short-cut method:

$$\frac{\begin{array}{l}3\text{ erect-eared} + 1\text{ droop-eared}\\ \times\ 3\text{ barkers}\quad + 1\text{ silent}\end{array}}{9\text{ erect barkers} + 3\text{ droop barkers} + 3\text{ erect silent} + 1\text{ droop silent}}$$

2. In minks either gene p or i, when homozygous, causes a platinum coat color. At least one dominant allele of each of these genes is necessary to produce the common brown color. Show the expected offspring of a cross between brown minks heterozygous for both genes. $(P/p\ I/i)$.

Answer. Using short-cut method:

$$\frac{\begin{array}{l}3\text{ brown }(P-) + 1\text{ platinum }(pp)\\ \times\ 3\text{ brown }(I-)\ + 1\text{ platinum }(ii)\end{array}}{\begin{array}{r}9\text{ brown }(P-I-) + 3\text{ platinum }(I-pp) + 3\text{ platinum }(P-ii)\\ + 1\text{ platinum }(ppii)\end{array}}$$

9 brown : 7 platinum

3. In man the dominant gene B causes blaze, a white forelock of hair. The recessive gene a causes albinism and the hair is white all over. Hence, albinism would be epistatic to blaze. A man with blaze had an albino mother. He marries an albino woman who had parents with normal hair color. Show the combinations of children which they may have and the proportions of each. (Note that since the gene for blaze is relatively rare we would assume that the man was heterozygous, for there is little likelihood that both parents would carry the gene.)

Answer. The genotype of the parents would be: Man, $B/b\ A/a$, woman, $b/b\ a/a$.

$$\frac{\begin{array}{l}1\text{ blaze}\quad + 1\text{ normal}\\ \times\ 1\text{ normal} + 1\text{ albino}\end{array}}{1\text{ blaze}\quad + 1\text{ normal} + 1\text{ albino} + 1\text{ albino}}$$

1 blaze : 1 normal : 2 albino

4. In the domestic fowl there are two pairs of genes affecting the comb. When homozygous for the recessive forms of both genes, the chicken will have a single (normal) comb. A dominant allele of one of these genes, P, causes a pea comb. A dominant allele of the other, R, causes a rose comb. However, if a chicken has at least one of both dominants, P– R–, the comb will be walnut—a result of the combined action of both dominant genes. Show the expected offspring of a cross between two walnut-combed chickens, both heterozygous for both genes.

Answer. The genotype of the parents would be: $P/p\ R/r$.

$$\frac{\begin{array}{l}3\text{ pea } + 1\text{ single}\\ \times\ 3\text{ rose } + 1\text{ single}\end{array}}{9\text{ walnut (pea-rose)} + 3\text{ rose } + 3\text{ pea } + 1\text{ single}}$$

5. The chestnut coat color of horses is due to a recessive gene, while the dominant allele results in black. The pacing gait is due to a recessive gene, whereas the dominant allele results in the trotting gait. Show the types of offspring which could result from a cross of a black trotter, heterozygous for both genes, crossed with a chestnut pacer.

Answer. Using the short-cut method:

$$\frac{\begin{array}{l}1\text{ black } + 1\text{ chestnut}\\ \times\ 1\text{ trotter } + 1\text{ pacer}\end{array}}{\begin{array}{l}1\text{ black trotter } + 1\text{ chestnut trotter } + 1\text{ black pacer}\\ \qquad\qquad\qquad\qquad\qquad\qquad\ \ + 1\text{ chestnut pacer}\end{array}}$$

6: PROBABILITY

Probability is a mathematical study of the operation of the laws of chance. It plays an important role in genetics because there are so many chance events that enter into the transmission of characteristics from one generation to another.

APPLICATIONS TO GENETICS

The segregation of genes in meiosis is subject to the same laws of chance as would be operating in the toss of coins, the roll of dice, or the shuffling of cards. These same laws again come into operation as the gametes unite when a new life is begun. If we know the parental genotype we can determine the expected ratios of the different kinds of offspring by the methods discussed in Chapters 4 and 5. But, we certainly know that the offspring are not always exactly distributed according to this ratio. In a test cross between a heterozygous black guinea pig and a white guinea pig we have an "expected" ratio of 1:1. Yet, this word "expected," as used in this sense, does not mean that we actually would expect to find exactly one half of the offspring to be black and one half white. If we obtain six offspring and four are black and two are white, we certainly would not consider this unusual. In fact, if we made many such crosses we might even find some with all white offspring. The expected ratio indicates what the laws of chance would predict. The obtained ratio may vary, within certain limitations, from this expected ratio without being a significant deviation.

This is understandable when we realize that ratios indicate the chance or possibility of the different types of offspring appearing at each birth. The first guinea pig conceived has a chance of one half of being white. If the animal is white, it does not influence the chance of the second animal being white—the second also has a chance of one half, and so on for all of the guinea pigs conceived.

We can correlate this with tossing coins. Suppose we have six coins to toss. The chance of obtaining a head is one half for each coin. If the first coin happens to be a head, this would not alter the chance that the second would also be a head, the second would not alter the chance of the third, and so on. If we tossed the six coins long enough we would eventually get six heads at one time. We can determine the frequency of such an occurrence by one of the simple laws of probability.

LAW OF COINCIDENT HAPPENINGS

Briefly this law may be stated as follows: ***The chance of any number of independent things happening together is equal to the product of the chances of each happening separately.*** If we toss two coins and want to know the chance of both coming up heads, we multiply the chance of a head for each: $\frac{1}{2} \times \frac{1}{2} = \frac{1}{4}$. This simply means that in one time out of four tosses, we would expect two heads. To find the chance of three heads in a toss of three coins, we would multiply by $\frac{1}{2}$ again and obtain $\frac{1}{8}$. For six coins we can see that we would expect all heads only once in a total of 64 tosses.

Thus while six white guinea pigs from heterozygous black parents would be very unusual, this result would occasionally be expected if the number of crosses were 64 or more.

Use in Human Heredity. To apply the principle to human heredity, let us determine the chance that heterozygous brown-eyed parents would have a blue-eyed girl as their first child. We know that the chance for blue eyes is $\frac{1}{4}$ and the chance for a girl is approximately $\frac{1}{2}$. We multiply these two fractions and find that the chance is only $\frac{1}{8}$ that the child will be both blue-eyed and a girl. If they wanted their first two children to be blue-eyed girls, they would have only one chance in 64 of satisfying their wishes ($\frac{1}{8} \times \frac{1}{8} = \frac{1}{64}$).

As another illustration, suppose a normally pigmented couple have an albino child as their first child. They had planned to have three more children and want to know the chance that all will be normal. The child already born would not enter into the calculations, since this is an event that has already occurred and will not influence events yet to come. Since the couple are, obviously, heterozygous for the gene for albinism, they would have a chance of ¾ of obtaining a normally pigmented child at any one birth. Hence the chance for three with normal pigmentation would be: $¾ \times ¾ \times ¾ = {}^{27}\!/_{64}$.

It is easy to see how probability used in this way can be of value in genetic counseling, especially where inherited abnormalities are involved.

Fig. 6.1. Coincident happenings. One can calculate the chance of obtaining two "ones" in a roll of a pair of dice by multiplying the chance of getting a "one" with one die. Since the chance of getting a "one" is one sixth then the chance of getting two "ones" is one thirty-sixth.

Effect of Previous Events. As brought out in the case of albinism, events which have already taken place do not influence events to come if the events are truly independent, as they should be if you are to employ the probability method. If you roll a pair of dice, the chance of obtaining two ones is $⅙ \times ⅙ = {}^1\!/_{36}$. But, suppose you roll them one at a time and you happen to get a one on the roll of the first. Now, the chance that you will obtain two ones is only ⅙ because the fact that the first roll gave a one will not influence the second die—it is just as likely to give a one as if the first had never been rolled.

In some cases the first event may have some influence on the chances of the second and due allowances must be made. For example, what are the chances of drawing two aces from a deck of cards? The chance of the first card being an ace would be $4\!/_{52}$, but if this card is an ace there are now only three aces in 51 cards,

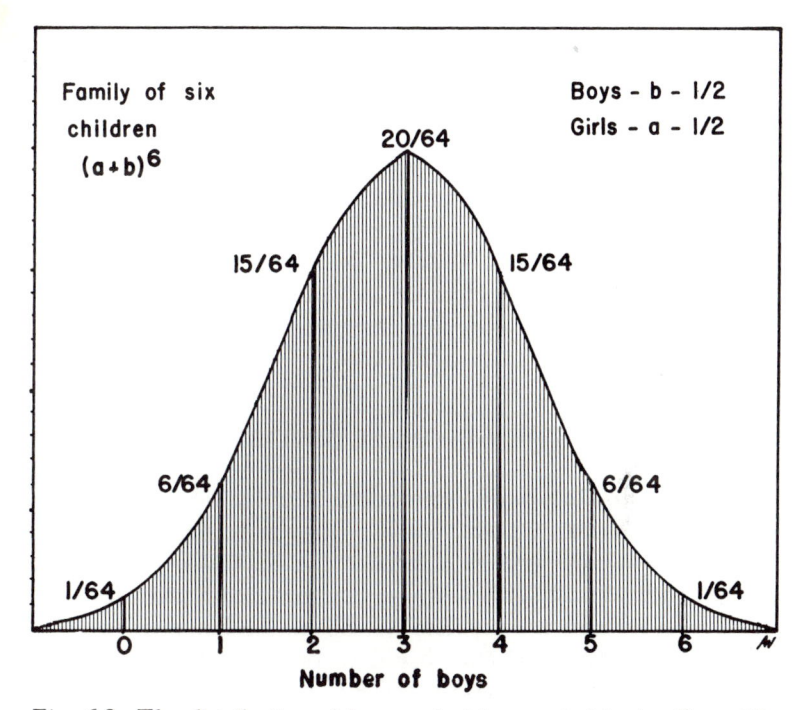

Fig. 6.2. The distribution of boys and girls expected in families with six children. This chart assumes that the chance of each sex is one half, although actually we know that there is a slight excess of girl babies which would alter the chart to a small extent if we were to use exact fractions.

so the chances of drawing an ace from such a deck is $3/51$. Hence, the chance of drawing two aces from one deck is $4/52 \times 3/51 = 12/2652$. This is an example of a case where the events are not entirely independent. We could make them independent, however, if we asked for the chance of drawing an ace from each of two separate decks of cards. Then the result would be $4/52 \times 4/52 = 16/2704$.

In fertilization the number of gametes is usually so large that the use of one gamete does not influence the chance that a similar gamete will be used in a second fertilization. For example, suppose a couple has just married and they would like to have four boys. Sex is determined by the type of sperm which fertilizes the egg. Sperm which will produce boys and those which will produce girls are produced in equal numbers and, if we assume that each has an equal chance at fertilization, then the chance of having a boy at any one birth would be $\frac{1}{2}$. About two hundred million sperms are released at one time and if one male-determining sperm were to be used in fertilization, this would not influence the chances of such a sperm's being used in future fertilizations. Hence, the answer to this problem would be $(\frac{1}{2})^4$ or $\frac{1}{16}$. If they already have three boys, however, the chance that they will have four boys would be $\frac{1}{2}$. Here we are dealing with the chance of a boy at only one birth—the first three do not in any way influence this chance.

THE BINOMIAL METHOD

The multiplication of the chances of independent happenings works well in many cases, but it is simpler to use the binomial method when we wish to determine the chances of various mixtures, such as three girls and one boy in a family with four children. To solve such a problem we expand the binomial $(a+b)^n$. We allow a to equal the chance of one event, b to equal the chance of the alternate event, and n to represent the total number of events considered. In this case the number of events would be 4 and the expanded binomial would be:

$$(a+b)^4 = a^4 + 4a^3b + 6a^2b^2 + 4ab^3 + b^4$$

We allow a to represent a girl with a chance of $\frac{1}{2}$ and b to represent a boy with a chance of $\frac{1}{2}$. The exponents represent the number of offspring. Hence we choose the second term in the expansion, $4\ a^3b$, to determine the chance of obtaining 3 girls and 1 boy in a total of 4 offspring. Our answer is: $4 \times (\frac{1}{2})^3 \times \frac{1}{2} = \frac{4}{16}$. Out of 16 marriages with four children, we would expect to find about 4 to have 3 girls and 1 boy.

If we worked out all of the possibilities for the entire binomial the results would be:

$\frac{1}{16}$ 4 girls: $\frac{4}{16}$ 3 girls 1 boy: $\frac{6}{16}$ 2 girls 2 boys: $\frac{4}{16}$ 1 girl 3 boys $\frac{1}{16}$ 4 boys.

The reader will note that this distribution forms a bell-shaped curve when plotted on a graph, which will always be the case

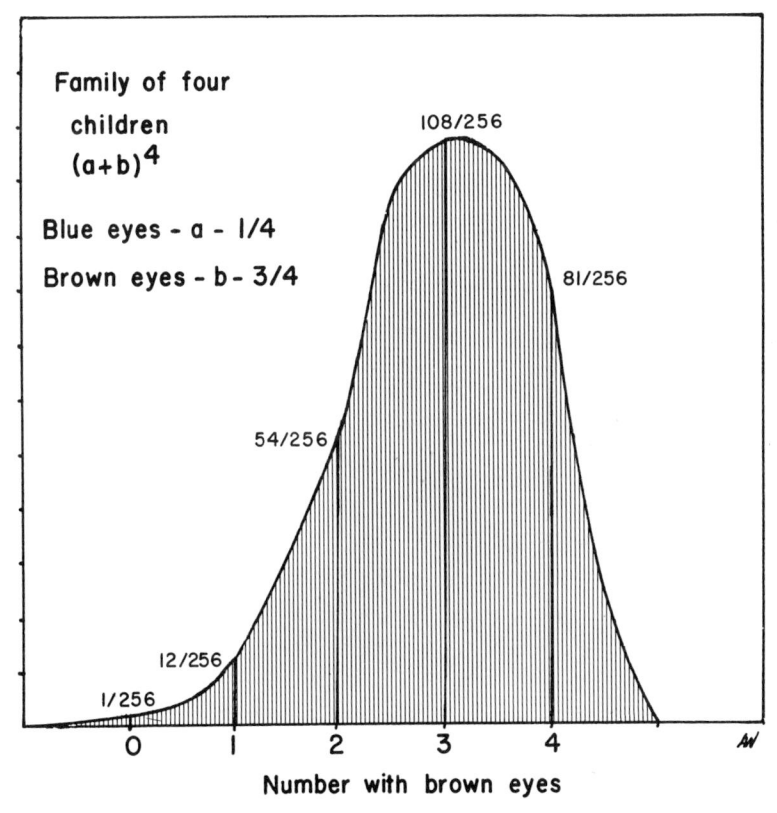

[Figure: Family of four children $(a+b)^4$. Blue eyes - a - 1/4. Brown eyes - b - 3/4. Distribution values: 1/256, 12/256, 54/256, 108/256, 81/256. X-axis: Number with brown eyes (0-4).]

Fig. 6.3. A skewed curve of distribution such as is found when the chance of two independent events are not equal. In this case the children expected from heterozygous brown-eyed parents are used as an illustration.

when the chance of each of both events is equal. Should the events not be of equal probability the curve will be skewed as shown in Fig. 6.3.

SHORT CUTS TO EXPANSION OF THE BINOMIAL

The binomial $(a+b)^n$ can be expanded by algebraic multiplication, but any specific expansion can be obtained more quickly by simple methods. Let us take $(a+b)^4$ as an example. The first term is always the first letter with an exponent the same as n. This would be a^4. The exponents of the second term would be a^3b. We can see that as we go through with the expansion the exponents of a drop one with each term and those of b increase one. (Note that the exponents always total 4.) In addition to the exponents, however, there are numbers—coefficients—preceding the terms. The coefficient of the first term is always 1 and is usually not put in. The coefficient of the second term is the same as the exponent of the first (in this case, 4). We can get the coefficient for the next term from the second term. We do this by multiplying the coefficient of this second term (4) by the exponent of the first

PASCAL'S TRIANGLE

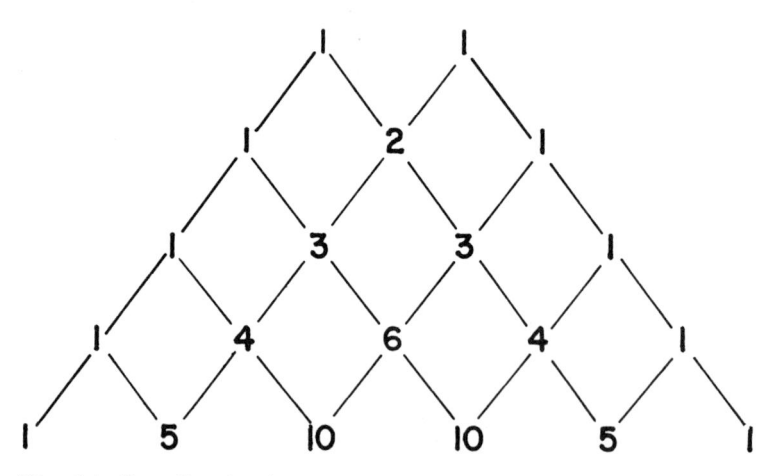

```
            1       1
          1     2     1
        1     3     3     1
      1     4     6     4     1
    1     5    10    10     5     1
```

Fig. 6.4. Pascal's triangle, a simple method of finding the coefficients in expanding the binomial $(a + b)^n$. The second number in each horizontal line represents the power of the binomial. Each coefficient is obtained by adding the two numbers above it.

letter (3) and dividing by the place in the expression. Since the place is the second term in the expression, this quotient would be 2. So, $4 \times 3 \div 2 = 6$, which is the coefficient for the third term. The coefficient of the next term would be $6 \times 2 \div 3 = 4$. For the final term it would be $4 \times 1 \div 4 = 1$.

Another simple method of obtaining coefficients involves the use of Pascal's triangle, as illustrated in Fig. 6.4.

STATISTICAL METHODS OF ANALYSIS OF RESULTS

When we compare the results obtained from a genetic cross with the results calculated by the laws of probability, we know that they will seldom appear in exact agreement. Almost always there will be some deviation due to chance variation. Sometimes, of course, there may be variations which are not due solely to chance. For instance, in 200 offspring of a cross between two normal *Drosophila* heterozygous for the gene for vestigial wings, we might obtain 156 offspring with normal wings and 44 with vestigial wings. By cursory examination we would notice that this deviation (6 of each class) is too small to be significant—this condition would be within the realm of chance variation from the 3:1 ratio. But, suppose we obtained 170 *Drosophila* with normal wings and 30 with vestigial wings. Perhaps this would be a deviation of sufficient size to indicate that something other than chance was operating—perhaps the gene for vestigial wings causes more larval and pupal deaths so that there is a lower survival rate when they are homozygous for this gene. This would appear to be so, but just where do we draw the line to indicate significant deviation? To answer such a question objectively, we must have some standard means to measure the deviation.

The Standard Error. One rather simple method of doing this is through calculation of the standard error. We obtain this by multiplying the expected number of one class of flies (150) by the expected number of the second class (50), dividing by the total number of flies (200), and extracting the square root of the figure obtained. By allowing p to stand for the first expectation, q for the second, and n for the total number, we obtain the following expression:

$$\text{SE} = \sqrt{\frac{p \cdot q}{n}}$$

(In some books one will find that the denominator is given as $n-1$, based on certain mathematical principles, and this procedure can be used if one prefers it. However, if the numbers involved are so small that the subtraction of one from the total will make any significant difference in the standard error obtained, they are too small to give an accurate standard error.)

After substitution of the proper numbers in our example, this expression becomes:

$$\text{SE} = \sqrt{\frac{150 \cdot 50}{200}} = 6.1 \qquad \frac{d}{\text{SE}} = \frac{6}{6.1} = .98$$

Now what does this standard error of 6.1 mean in relation to the obtained deviation of 6? It means that if many crosses of this nature were made and the 200 offspring were examined, we would obtain a deviation as great as, or greater than the SE in about ⅓ of the cases (32 per cent). The deviation obtained (6) is almost exactly equal to the SE (6.1). Since a deviation of more than 6.1 would be expected in about ⅓ of the cases, we would certainly have no reason to think that anything other than pure chance caused our deviation to be this great. In general, we say that a standard deviation must be at least twice the SE before it is significant.

Let us now see whether there would be a significant deviation if the results observed were 170 black offspring and 30 white off-

Deviations of Different Magnitude Relative to the Standard Error

$\frac{d}{\text{SE}}$	Per Cent of Cases with Deviations This Great or Greater	Odds Against Occurrence of Deviation This Great or Greater
.6745	50.00	1.00:1
1.0	31.73	2.15:1
1.3	19.36	4.17:1
1.6	10.96	8.12:1
1.8	7.19	12.92:1
2.0 (sig.)	4.55	20.98:1
2.2	2.78	34.96:1
2.4	1.64	60.00:1
2.6 (highly sig.)	0.932	106.30:1
2.8	0.511	194.70:1
3.0	0.270	369.40:1
3.5	0.0465	2,149.00:1
4.0	0.00634	15,770.00:1
5.0	0.0000573	1,744,000.00:1
6.0	0.00000020	500,000,000.00:1

spring. This would be a deviation of 20 of each class from the expected 3:1 ratio. The SE, of course, remains the same since the total number of flies is the same.

$$\frac{20}{6.1} = 3.3$$

Since the obtained deviation is 3.3 times the SE, we conclude that this is a highly significant deviation and we have to seek causes other than chance to explain so great a deviation.

It is sometimes valuable to know in just what percentage of the cases will be found a deviation of a certain magnitude. The preceding table shows this percentage for certain deviations and also indicates the odds against occurrence of deviations of various magnitudes.

The Use of Percentages. The standard deviation can also be calculated by employing percentages of the deviations. We can use the same results of the *Drosophila* cross to illustrate this procedure.

Total number of offspring	(n)	200
Expected per cent—normal wings	(p)	75
—vestigial wings	(q)	25
Obtained per cent—normal wings		78
—vestigial wings		22
Deviation of both from expected	(d)	3

$$SE = \sqrt{\frac{p \cdot q}{n}} = \frac{75 \cdot 25}{200} = 3.07\%$$

$$\frac{d}{SE} = \frac{3}{3.07} = .98$$

We can see that the percentage methods results in the same answer as the one worked out with numbers. The choice of methods would depend upon the convenience of handling the figures involved.

The Use of Chi-square. Another valuable tool which may be used in determining the "goodness of fit" of the results of genetic crosses is known as chi-square. It is simply the sum (Σ) of the squares of the deviations (d) divided by the expected numbers (e). It can be represented in condensed form as:

$$\chi^2 = \sum \left(\frac{d^2}{e}\right)$$

An example presented in tabular form can be employed to illustrate the chi-square procedure. We will designate the second set of numbers of offspring from *Drosophila* heterozygous for vestigial wings, 170:30, which proved to be a significant deviation when analyzed by the standard error.

F_2 Offspring	Observed x	Expected e	Deviation Squared d^2	$\dfrac{d^2}{e}$
Normal	170	150	400	2.66
Vestigial	30	50	400	8.00
	200			$\chi^2 = 10.66$

P < 1%

Chi-square comes out as 10.66. How then do we interpret this figure? Whenever we are dealing with two classes, any figure that comes out higher than 3.8 is considered significant, for 3.8 indicates the level where chance occurrence could account for deviations as great as this in only 5 per cent of the cases. Since the chi-square obtained is definitely higher than this number, we would say that it represents a significant deviation.

Degrees of Freedom	Possibility of Chance Occurrence in Percentage (5% or Less Considered Significant)								
	90%	80%	70%	50%	30%	20%	10%	5% (sig.)	1%
1	0.016	0.064	0.148	0.455	1.074	1.642	2.706	3.841	6.635
2	0.211	0.446	0.713	1.386	2.408	3.219	4.605	5.991	9.210
3	0.584	1.005	1.424	2.366	3.665	4.642	6.251	7.815	11.341
4	1.064	1.649	2.195	3.357	4.878	5.989	7.779	9.488	13.277
5	1.610	2.343	3.000	4.351	6.064	7.289	9.236	11.070	15.086
6	2.204	3.070	3.828	5.348	7.231	8.558	10.645	12.592	16.812
7	2.833	3.822	4.671	6.346	8.383	9.803	12.017	14.067	18.475
8	3.490	4.594	5.527	7.344	9.524	11.030	13.362	15.507	20.090
9	4.168	5.380	6.393	8.343	10.656	12.242	14.684	16.919	21.666

We can determine more closely how often deviations of this magnitude would be expected to occur by reference to a table of chi-square values. The above table shows that the chi-square values vary according to the degrees of freedom which pertain to the problem. With two classes we have one degree of freedom—the degree of freedom is the total number (n) minus one. As we look along this column of figures we find that our obtained chi-square value is greater than the value given for a 1 per cent chance occurrence. Hence, we can say that P (probability of such an occurrence by chance) is less than 1 per cent ($P < 1\%$) and, therefore, is highly significant. This bears out the findings obtained through use of the standard error on these same figures.

A more detailed explanation of what is meant by degrees of freedom might help in the use of chi-square. Suppose you are the judge of a beauty contest. Four contestants are entered and you are to make four awards. You have freedom to award first prize to any one of the four. Having made this award you still have freedom in choosing the second place winner. For the third prize you have freedom of choice among the two remaining, but for the fourth prize you have no freedom of choice—it must go to the remaining contestant. Hence, you have only three degrees of freedom in judging the four girls ($n - 1$).

As applied to a genetic problem, there is a degree of freedom in the number of vestigial-winged flies obtained among 200 offspring, but if this number turns out to be 30 there is no freedom in the number that do not have vestigial wings—this must be the remainder of 170.

As an example of the use of chi-square for the results of a dihybrid cross, let us take the actual results obtained by Mendel in one of his garden pea crosses.

Phenotype of F_2	Observed x	Expected e	Deviation Squared d^2	$\dfrac{d^2}{e}$
Yellow-round	315	313	4	.0128
Yellow-wrinkled	101	104	9	.0865
Green-round	108	104	16	.1538
Green-wrinkled	32	35	9	.2571
	556			$\chi^2 = .5102$

Upper limits of chi-square at 5% level of significance 7.815
$$P > 90\%$$

When we compare the chi-square value of .5103 with the line for three degrees of freedom in the table we find that deviations of this magnitude would occur in over 90 per cent of the cases ($P > 90\%$).

Usually it is sufficient to express the value of P in this way; if it lies between 20 and 30 per cent, we may express it as $30\% > P > 20\%$. Or, if more exact figures are needed, we can obtain a close approximation by extrapolation. Thus, a chi-square of 1.61 for one degree of freedom would give a value of $P = 20.56$ per cent.

Two precautions should be kept in mind about the use of chi-square. *First*, it cannot be used with percentages—the actual numbers obtained must be used. *Second*, the results become less accurate when the number of expected events in any class becomes small and it will not be reliable when the number is less than five for any one class.

TYPICAL PROBLEMS AND ANSWERS

1. Suppose you carry a gene for a certain rare blood type. What is the chance that a specific great-grandchild of yours will receive this gene from you?

Answer. The chance of any one child receiving the gene from you is one half; since there are three generations, the answer is:

$$\tfrac{1}{2} \times \tfrac{1}{2} \times \tfrac{1}{2} = \tfrac{1}{8}$$

2. In the United States about one Negro out of every 576 has sickle-cell anemia which can be recognized by the sickle shape assumed by the red cells when they are in a low oxygen concentration. The disease is practically unknown among whites. The anemia is found when a person is homozygous for a certain recessive gene that is concerned with hemoglobin formation. In a certain southern city about one out of every four persons is a Negro. What are the chances that a person chosen at random in this city will have sickle-cell anemia?

Answer.

$$\tfrac{1}{576} \times \tfrac{1}{4} = \tfrac{1}{2304}$$

3. About 10 per cent of the people in the United States have type B blood. A wedding picture in the newspaper shows a newly married couple. What are the chances that both will have type B blood?

Answer.

$$\tfrac{1}{10} \times \tfrac{1}{10} = \tfrac{1}{100}$$

TYPICAL PROBLEMS AND ANSWERS 83

4. If the couple both have type B blood, what are the chances that their first child will have type O blood? (About 3/7 of the people with type B blood are heterozygous for the recessive gene which produces type O blood.)

Answer. First obtain the chance that two type B persons will both be heterozygous. $3/7 \times 3/7 = 9/49$. Since the chance that heterozygous parents will have a child with a recessive trait is 1/4, we multiply by this fraction.

$$9/49 \times 1/4 = 9/196$$

5. The tendency to have the disease of hemolytic jaundice is due to a dominant gene, but only 10 per cent of the people with the tendency actually develop the disease. A heterozygous man marries a homozygous normal woman. Show the proportion of the children that would be expected to develop hemolytic jaundice.

Answer. The chance that a child will receive the gene from the father is 1/2 and the chance of showing the disease if he gets the gene is 1/10. Thus:

$$1/2 \times 1/10 = 1/20$$

6. A newly married couple plan on having four children and would like to have two of each sex. What are the chances that their wishes will be fulfilled?

Answer. Assuming the chance of each sex as 1/2, we proceed:

a = chance of girl = 1/2 b = chance of boy = 1/2

From binomial $(a+b)^4$ we choose $6a^2b^2$ $6(1/2)^2(1/2)^2 = 6/16$

7. A certain couple with normal pigmentation have two albinos and one normal child. How often would you expect such a combination to appear?

Answer. Parents are obviously heterozygous for the condition; and thus:

a = chance of albino = 1/4 b = chance of normal = 3/4

From binomial $(a+b)^3$ we choose $3a^2b$ $3(1/4)^2(3/4) = 9/16$

8. A man has a roan bull and a white cow. His first 5 calves turn out to be 2 roans and 3 whites. Keeping in mind the fact that roan is a color obtained when cattle are heterozygous for the genes for red and white, how frequently would you expect such a combination of offspring as that obtained in this case?

84   PROBABILITY

Answer.

$$a = \text{roan} = \tfrac{1}{2} \qquad\qquad b = \text{white} = \tfrac{1}{2}$$

From binomial $(a + b)^5$ we choose $10\, a^2 b^3 = {}^{10}\!/_{32}$

9. Expand the binomial $(a + b)^6$ using the short-cut method.

Answer.

$$(a + b)^6 = a^6 + \underset{6\,a^5 b}{(1 \times 6 \div 1)} + \underset{15\,a^4 b^2}{(6 \times 5 \div 2)} + \underset{20\,a^3 b^3}{(15 \times 4 \div 3)}$$
$$+ \underset{15\,a^2 b^4}{(20 \times 3 \div 4)} + \underset{6\,a b^5}{(15 \times 2 \div 5)} + \underset{b^6}{(6 \times 1 \div 6)}$$

10. Let us say that statistics show that cancer contributes to death in about $^1\!/_{10}$ of the U. S. population. There is considerable interest in the part which heredity might play in cancer. You find that your family's pedigree, as far as you can trace it, includes 64 persons who are dead; 13 of these died of cancer. Is this a significant deviation from the expected?

Answer.

Total number (n)	64.0
Expected—cancer (p)	6.4
—not cancer (q)	57.6
Obtained—cancer	13.0
—not cancer	51.0
Deviation (d)	6.6

$$\text{SE} = \sqrt{\frac{6.4 \cdot 57.6}{64}} = 2.4 \qquad \frac{d}{\text{SE}} = \frac{6.6}{2.4} = 2.8$$

Since the deviation in this pedigree (6.6) is 2.8 times its SE, we would call this highly significant. According to the table on probabilities for SE, such a deviation would occur by chance only once in about 200 such studies. This would support the concept that heredity might play a part in causing cancer, but other factors involved would certainly have to be analyzed before arriving at any conclusions.

11. There is also interest in the possible relation of heredity to the site of cancer. Suppose statistics show that $^1\!/_3$ of all cases of cancer in women are breast cancer. In the pedigree mentioned in Problem 9, suppose you find that (1) 9 women died of cancer and (2) 6 of these had breast cancer. Is this a significant deviation from the expected number?

Answer.

Total number of women (n)	9
Expected—breast cancer (p)	3
—other sites (q)	6
Obtained—breast cancer	6
—other sites	3
Deviation (d)	3

$$SE = \sqrt{\frac{3 \cdot 6}{9}} = 1.4 \qquad \frac{d}{SE} = \frac{3}{1.4} = 2.1$$

Since this deviation is slightly more than twice the standard error, we would say that this is a significant deviation and again might suggest that heredity plays a part in the site of cancer. With such a small number however, the matter is open to question and would require further tests for confirmation.

12. In an isolated group of 228 Pennsylvania Germans known as the Dunkers, Dr. Bentley Glass of Johns Hopkins University found about 60 per cent with type A blood. The general United States white population has about 40 per cent type A. Is this a significant deviation?

Answer.

Total number of people (n)	228
Expected per cent—type A (p)	40
—other types (q)	60
Observed per cent—type A	60
—other types	40
Deviation (d)	20

$$SE = \sqrt{\frac{40 \cdot 60}{228}} = 3.24 \qquad \frac{d}{SE} = \frac{20}{3.92} = 6.1$$

This is certainly a significant deviation and thus indicates that there must be something other than chance which caused such a high proportion of type A individuals in this group. Genetic drift seems to be the reason.

7: THE DETERMINATION OF SEX

Sex differentiation between the male and female of higher animals is said to be very great. The differences are far too extensive to be accounted for by the principles of single gene transmission. In human beings, for example, the distinctions extend far beyond those obvious characteristics which are associated with reproduction. The skin, muscles, blood, bone, and hair show differences—in fact, there is hardly any part of the body that is not affected in some way by the sex of the individual.

SEXUAL BIPOTENTIALITY OF ORGANISMS

The problem of a method of determination of sex becomes much simpler if, from the beginning, we recognize a basic fact. All organisms characterized by sexual reproduction appear to have the potentialities of both sexes in their genic make-up. The most masculine man has genes within his body which can form all of the characteristics associated with a woman. In many of the simpler forms of animal life and in most forms of plant life, both male and female characteristics are expressed in all organisms. Usually, there is some means for cross-fertilization, but all individuals can serve as both male and female in the reproductive process.

THE SEPARATION OF THE SEXES

Among most forms of higher animals there are two separate sexes with various degrees of differentiation of the body parts. It is here that the problem of determination of sex becomes evident. Which of the animals shall express the female characteristics inherent in its genes and which shall express the male? There are a few cases among the simpler forms of animal life wherein sex is determined by environment. In the marine worm, *Bonelia*, for example, all newly hatched worms may become either male or female. If they are reared in relative isolation from other worms, they will *always* become females. On the other hand, if the newly hatched worms find themselves near mature females, some of them will attach themselves to the proboscis of a female and these become males. The males are much smaller than the females and become parasites upon them—actually living within the genital tract of the females. Apparently, there is something in the secretions of the proboscis of the female that causes the genes for maleness to be expressed and the genes for femaleness to be suppressed.

This method is not the most efficient, for it tends to produce varying numbers of the two sexes and it is not surprising that we find it limited to a very few animals.

SEX DETERMINATION BY CHROMOSOMES

Chromosomes are involved in sex determination of the great majority of forms of life that have separate sexes. Since the chromosomes follow a definite and predictable pattern of segregation in meiosis, they would be expected to yield a much more reliable distribution of the sexes.

Discovery of Sex Chromosomes. In the early days of cytological investigation (1902), C. E. McClung noted an odd chromosome in the diploid cells of the male grasshopper. There were eleven paired chromosomes and this extra one. He called it the *X-chromosome* (X for unknown) and proposed the theory that it was related to sex determination. Later (1905), N. M. Stevens, of Columbia University, found that the females had this chromosome also, but there was a pair of them in each diploid cell. Further investigation showed that in the males of some species the X-chromosome in meiosis became paired with a chromosome of a different size and shape. This became known as the *Y-chromosome.* As the relation of these two chromosomes to sex determination became better understood they were also designated as *sex chromosomes.* Then, as a matter of convenience in discussing sex determination, the other chromosomes in the cell were designated as *autosomes.*

The XY Method. Since the XY method is the most common method of sex determination we will discuss it first, using *Drosophila melanogaster* as an example. The diploid chromosome

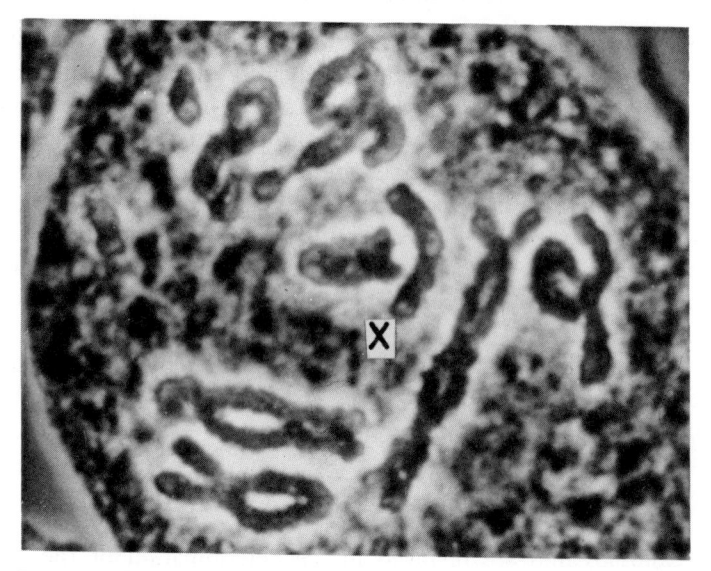

Fig. 7.1. The X-chromosome of the grasshopper can be plainly seen in this photomicrograph of a smear of the testes. This cell is a primary spermatocyte. The autosomes are all in pairs, but the single X-chromosome has no partner. It was this observation which led to the discovery of the chromosome method of sex determination.

number of this insect is eight. There are three pairs of autosomes and one pair of X-chromosomes in the female diploid cells and three pairs of autosomes and the paired X- and Y-chromosomes in the male. The sex chromosomes are easily distinguished from one another because the X-chromosome is long and straight while the Y-chromosome is bent and J-shaped.

In oögenesis the chromosomes pair and the eggs and polar bodies all receive the same kind of chromosomes—three autosomes and one X-chromosome. This is not the case in spermatogenesis, however, for one pair of chromosomes is unlike and half of the sperms receive the X-chromosome while the other half receive the Y-chromosome. Sex determination, therefore, depends upon which of the two types of sperms fertilizes the egg. Hence, we can think of the sperms carrying the X-chromosome as being female-determining while those carrying the Y-chromosome are male-determining.

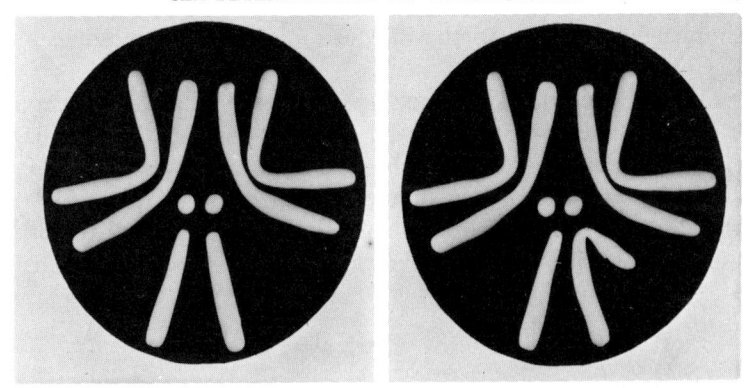

Fig. 7.2. Chromosome differences of the sexes in Drosophila melanogaster. *These clay models show the female cell on the left with two X-chromosomes at the lower part of the cell. The male has one X-chromosome and a Y-chromosome which is characterized by a hook on one end.*

The majority of *dioecious* animals (those with separate sexes) have the XY method of sex determination. This majority includes man. In most of these animals the Y-chromosome is considerably smaller than the X-chromosome. Also, the number of genes in the Y-chromosome is always much smaller than the number present in the X-chromosome. There are a few genes which the Y shares in common with the X, and these are sufficient to bring about pairing during meiosis. Furthermore, there appear to be a few genes on the Y-chromosome which have no homologous mates on the X-chromosome. (See Chapter 8 for further discussion.)

The XO Method. The XO method of sex determination is found in many *Orthoptera* (grasshoppers, etc.) and *Heteroptera* (bugs). Since the Y-chromosome plays no part in sex determination in *Drosophila*, we would not expect to find that its absence alters the method of sex determination in other insects. It does not; the method is the same except for the fact that there exists no Y-chromosome.

This creates an unusual condition in the diploid chromosome number. A typical grasshopper female has 24 $(22 + XX)$ as the diploid chromosome number, but a male has only 23 $(22 + X)$. All eggs carry 12 chromosomes, but there are two kinds of sperm—one carrying 12 (female-determining), the other carrying 11 (male-determining) chromosomes. It was this unusual chromosome

90 THE DETERMINATION OF SEX

complement that led to the discovery of the X-chromosome by McClung.

The ZW Method. There are some animals in which the relation of the sex chromosomes is exactly reversed—that is, the male has all paired chromosomes, while the female has one unmatched pair. Thus, the eggs are of two different kinds; sex is determined by which type of egg is fertilized since the sperm are all alike with respect to sex chromosomes. The chromosomes may be designated

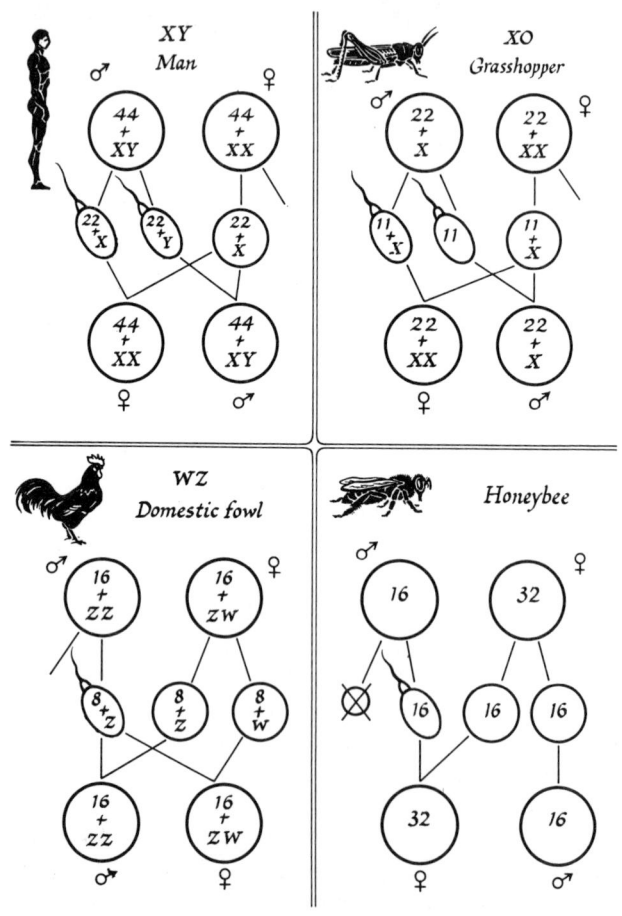

Fig. 7.3. The four major types of sex determination by chromosomes in animals. (From Winchester, Genetics, Houghton Mifflin.)

as X- and Y-chromosomes, but to avoid confusion we sometimes use **Z-chromosome** for the equivalent of the X and **W-chromosome** for the equivalent of the Y. Thus, the male has the ZZ combination, whereas the female has the ZW combination. Butterflies, moths, caddis-flies, birds, and some fishes are known to have this method of sex determination.

The Honeybee Method. There is an unusual type of sex determination in the honeybee and certain other *Hymenoptera*. The males are haploid with respect to all chromosomes, but the females are diploid. Eggs are produced by normal oögenesis and, therefore, have the haploid number of chromosomes. Sperms are produced by a special type of meiosis which allows all of the chromosomes to go to one cell and none to the other. This type of union yields haploid sperms.

Females, such as the queen bee, are inseminated during the mating process, but retain the sperms within a seminal receptacle and can lay either fertile or infertile eggs. They can constrict the duct coming from the seminal receptacle and prevent sperms from coming down as an egg is laid. This gives an infertile egg which, being haploid, will hatch into a male. Fertilized eggs, on the other hand, are diploid and become females. This allows the female a means of control over the sex of the offspring which is not possessed by animals having any other method of sex determination. In the honeybee it enables the queen to lay fertile eggs in the smaller cells for worker bees (females) and infertile eggs in the slightly larger cells which accommodate the drones (males).

HORMONES AND SEX DETERMINATION

In vertebrate animals the gonads produce secretions known as sex hormones which play an important part in the development of sexual characteristics. If the testes are removed from a preadolescent boy, he never develops the characteristics commonly associated with the male sex. The voice remains high pitched, the beard does not develop, the muscles do not become masculine, and the normal male interest in the opposite sex does not develop. Castration of an immature female, likewise, produces a person who fails to develop the normal female sexual traits.

Sex Reversal. Experimentally, it is possible to achieve a great degree of reversal of the genetic sex through castration and ad-

Fig. 7.4. Experiments to demonstrate hormone effects in the domestic fowl. A. Normal female. B. Female with ovaries removed—cock feathering has developed. C. Female with ovaries removed and testes engrafted—typical male characteristics have developed. D. Female with engrafted testes and ovary not removed. (From Winchester, Genetics, *Houghton Mifflin.)*

ministration of the hormones of the opposite sex. In chickens, for instance, the ovaries can be removed from a young female, and testes from a male chick can then be implanted. These testes will grow and will release the male hormone. Upon maturity the bird is indistinguishable from a rooster that has the normal chromosome complement for a male. The altered bird, however, will be sterile because there will not be the proper connection to the ducts which are necessary routes for the conduction of sperms.

Sometimes there is a spontaneous partial reversal among old hens and they begin to crow, develop male feathers, and even mate with other hens. Studies on such hens show that disease has destroyed the normal ovarian tissue and there is a bit of rudimentary testicular tissue which begins growing and produces the male hormone. There are a few isolated cases in which such reversed females have produced motile sperms and fathered chicks.

Among mammals such gonad transplantation is very difficult because of the antigen-antibody reactions which develop when tissue from one body is ingrafted in another. It is possible, however, to achieve a high degree of reversal by removal of the gonads and the administration of the hormones of the opposite sex. At the present writing there are two widely known cases in which this operation has been performed on human beings. The degree

Fig. 7.5. Sex reversal? This widely publicized case of Christine represents an attempt at sex reversal through gonad removal and administration of the hormones of the opposite sex. The small insert shows this person as a man before the operations and the larger picture shows his appearance several years later. (Small photo, United Press.)

of reversal depends upon the degree of maturity that exists when the work is started. Characteristics which have already developed, such as masculine voice and beard in man, will not disappear when the hormones are switched.

Sex Chromosomes and Hormones. How can we correlate this information about the effects of hormones with the known facts about chromosome determination of sex? In the vertebrate animals it would appear that (1) the chromosome complex determines which of the hormones shall be secreted and (2) the hormones accomplish the balance of sex differentiation. The gonads of early embryos have two types of tissue—the outer layer which is ovarian in nature and the inner layer which is testicular in nature. Development proceeds along parallel lines for a time in both sexes; then one type of tissue in the gonads assumes ascendancy over the other and sex differentiation begins. Once sexual development has become well established, the chromosomes apparently cease their influence and the hormones carry on with the differentiation and maintenance of the male or female sex.

SEX INTERGRADES

There are times when various mixtures of male and female characteristics may occur in animals which normally have separate sexes because of various abnormalities of chromosomes or hormones.

Pseudohermaphrodites. In many of the lower forms of animal life it is quite common to find the sex organs of both sexes in each animal. The earthworm, for example, has both, and each earthworm can serve as both male and female in the reproductive process. They copulate (sexual union), but during this process each worm transfers sperms to the body of the other, and thus both serve as males. Later, each worm serves as a female and lays eggs which are fertilized by the stored sperm from the other worm. Animals which can serve both functions are called *hermaphrodites.*

Animal breeders know from experience that even among mammals there are rare cases in which both sexes are well developed in one body. These are abnormal and cannot function as either sex, so they are called *pseudohermaphrodites.* One theory to explain this condition, as postulated by Crew, holds that there may exist quickly elaborating female-determining genes which are able to trigger the enlargement of ovarian tissue in the anterior part of the embryonic gonads—even though the animal has the chromosome complex for a male. Later, the effect of the chromosomes causes the male-determining genes to begin functioning, and the posterior part of the gonads undergo development of the testicular tissue. Both hormones are produced and there is a partial development of both sets of sex characteristics. There are even a few substantiated cases of human pseudohermaphrodites.

Gynandromorphs. Among the animals that do not have sex hormones, it is possible to discover sex intergrades, with distinct areas of the body showing male and female tissue. These are known as gynandromorphs, a term frequently shortened to gynanders. The most striking of these is the bilateral sex mosaic. Such individuals are frequently found in *Drosophila* studies in which large numbers of flies are examined and especially at times when the females have been treated with radiation.

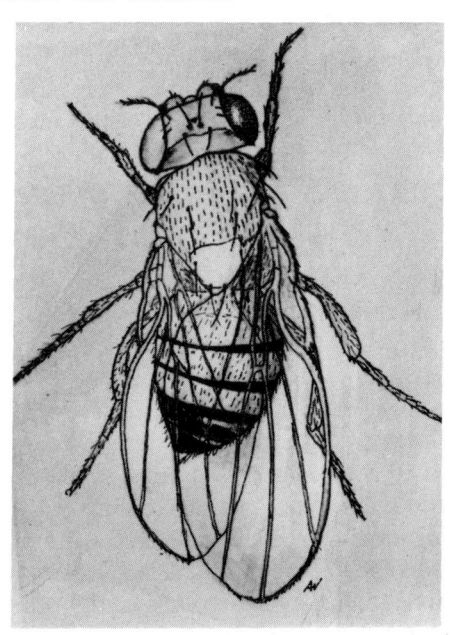

Fig. 7.6. Drosophila gynander. *This bilateral sex mosaic is male on the left side and female on the right. The left side is smaller because males are smaller than females. This fly was heterozygous for white eyes and the chromosome having the normal allele for white was lost on the left side so the eye on the male side is white.*

Gynanders apparently begin with the female XX chromosome complement, but during the first mitosis of the zygote one of the X-chromosomes lags behind and fails to be included in the nucleus of one cell. Lost in the cytoplasm, it disintegrates and plays no part in sex determination. This condition results in two cells, one with the ratio for male and one with the ratio for female traits. One forms one side of the body and one forms the other. Should the lag occur during the second mitosis of one of the two cells, a gynander would be produced with one fourth of the body showing male tissue, and even smaller areas of this kind can result when the lag occurs in later divisions.

Since hormones tend to affect all body areas equally, gynandromorphs would appear to be impossible among the vertebrates; although there was one supposed case reported from Russia in which there were scattered areas of tissue of both sexes on one person, it has been impossible to check the accuracy of this report.

VARIATIONS IN THE SEX RATIO

In all the methods of sex determination described, with the exception of the environmental method and the honeybee method, the male- and female-determining gametes are produced in equal numbers and, other things being equal, we would expect the sexes in a 1:1 ratio.

In Drosophila. Counts of the two sexes in *Drosophila* run very close to this ratio when large numbers are enumerated, although the females tend to come out of the pupa earlier and counts of flies from a bottle during the first day of emergence will almost always yield a significant excess of females. Within a few days, however, the numbers are balanced. In certain stocks there may be a consistent predominance of one sex or the other due to inherited lethals which affect one sex, but not the other.

In Man. There are certain human families in which the large number of one sex or the other would appear to be due to some genetic factor, and it is possible that this type of situation could be akin to the lethals in *Drosophila*, but scientists believe that the former condition simply indicates variations due to the laws of chance. In large families we would expect at least a few cases out of many thousands studied in which all the members would be of the one sex.

When we study the sex of live births over the entire United States, however, we find an excess of males over females that could not be due to chance. About 106 boys are born for every 100 girls. A possible explanation for this situation could be that there are more embryonic deaths among the girls; thus this prevailing condition would account for the discrepancy. About 16 to 20 per cent of all conceptions terminate in death before or at birth—perhaps most of these could be girl embryos. This possibility is ruled out, however, by analysis of the sex of the embryos that have failed to withstand the hazards of embryonic existence. Among the great collection of such embryos at the Carnegie Institution of Washington there are more males than females at all ages old enough to disclose sex. Hence, we must conclude that more male zygotes are formed.

How cán this be true, since human sperms are produced in equal numbers? One theory holds that, since Y-chromosome is much smaller than the X-chromosome, the sperms carrying the Y have a lighter weight and slightly smaller size that gives them a slight advantage in the race for the egg, or perhaps an advantage in penetrating the egg once they have arrived.

It is interesting to note that, in spite of the excess of males among babies, the sexes are nearly equal at about 20 years of age. In middle age the greater attrition among the males has continued. Women outnumber men about 100:85 at age 50. At 85 years of age there are about twice as many women as men. Hence, it is clear that, constitutionally, it is the women who are the stronger sex.

We see further evidence of this in studies of the sex ratio among babies born in other countries. In those countries where living standards are highest and an embryo would have the greatest chance of survival, the proportion of males to females is highest. Contrarily, in the regions where crowded living conditions, poor nutrition, and insufficient medical care would tend to reduce the chances of embryonic survival, we find a relatively low ratio of males to females.

Also, it has been found that the ratio of boys to girls in multiple births is lower than in single births. We would expect more embryonic deaths in multiple births and consequently more girls among the survivors. For twins in the United States the ratio is 103.5 boys to each 100 girls as compared to the ratio 106:100 when all births are included. For triplets it drops to 98:100 and for quadruplets the ratio stands at about 70:100.

ATTEMPTS AT PREDETERMINATION OF SEX

Many efforts have been made to control the sex of offspring in man and in other forms of animal life. Most have had no chance of success because they have been administered after conception. Different kinds of diets and all sorts of chemical concoctions have been tried on pregnant women in an effort to predetermine the sex of the child. Some recent attempts have had at least some chance of success because they have recognized the known facts of sex determination.

An ultracentrifuge has been tried in an effort to separate the X and Y sperms on the basis of a possible slight difference in weight. Also, efforts have been made to detect possible antigenic differences in the two in the hope that perhaps one or the other might be immobilized. The most promising work, however, appears to be in progress with the use of electrophoresis. It has been found that sperms suspended in a normal saline solution will migrate to one pole or the other when exposed to weak electric currents. The amount of current used determines to which pole they migrate. At a critical level near the point where there is a shift in attraction, some sperms go to one pole and some to the other. Experiments on rabbits indicate that sperms taken from one pole yield a significant excess of male offspring, whereas those taken from the other pole yield a significant excess of female offspring.

TYPICAL PROBLEMS AND ANSWERS

1. It has been found that injection of male hormone into the air space of incubating eggs of the domestic fowl will cause an excess of males to hatch. Apparently there has been a reversal of sex in some of the females. What unusual sex ratios might be expected from the descendants of these chickens?

Answer. Some of the males would have the ZW chromosome combination. When these are crossed with the normal ZW hens, we will get the following results:

$$ZW \times ZW = ZZ + 2\,ZW + WW$$

The WW die; hence, of living chicks hatched, there will be twice as many females as males.

2. Records show that there has been an increase in the ratio of males to females in live births in the United States during the past 30 years. What facts learned in this chapter might have a bearing on this?

Answer. With the great progress that has been made in prenatal care, there has doubtless been a great reduction in the prenatal deaths. This result would tend to increase the number of males in the living offspring since the males have a lower survival chance.

3. Studies in Holland show a sex ratio of 105.10 males to each 100 females among the legitimate births, but this drops to 103.63:100 among the illegitimate births. Can you suggest a possible explanation for this difference?

Answer. It is quite natural that unwed expectant mothers would not have the prenatal care that would be given those women undergoing pregnancy under more acceptable circumstances. This situation would result in a greater number of embryonic deaths among those not married and less chance for survival of the male embryos in comparison with the female.

8: RELATION OF SEX TO INHERITANCE

The differences in chromosomes and hormones which play a part in the determination of sex also have some bearing on the pattern of inheritance of many genes. It will be the purpose of this chapter to show what these effects can be.

SEX-LINKED GENES

Genes which lie on the X-chromosome (or the Z-chromosome) are said to be sex-linked because the pattern of transmission of these genes is related to sex determination and is somewhat different from the usual pattern of transmission as observed in the genes on the autosomes.

Discovery of Sex-linkage. In the early days of the studies of *Drosophila* at Columbia University, literally millions of flies were examined by T. H. Morgan and his students. Occasionally, the

Fig. 8.1. White-eyed, miniature-winged Drosophila. These two characteristics result from the action of two sex-linked genes.

researchers would find flies which showed inherited deviations from the normal or wild type, deviations which arose as mutations of the existing genes. One of these mutations caused a fly to have white eyes instead of the normal red. Through inbreeding the researchers soon obtained several of these white-eyed flies and began to notice that there existed an unusual method of inheritance for this characteristic.

A white-eyed female mated with a red-eyed male gave white-eyed male offspring and red-eyed female offspring. Morgan reasoned correctly that the gene for white eyes must be on the X-chromosome for this unusual method of inheritance to occur. The male receives its single X-chromosome from the female

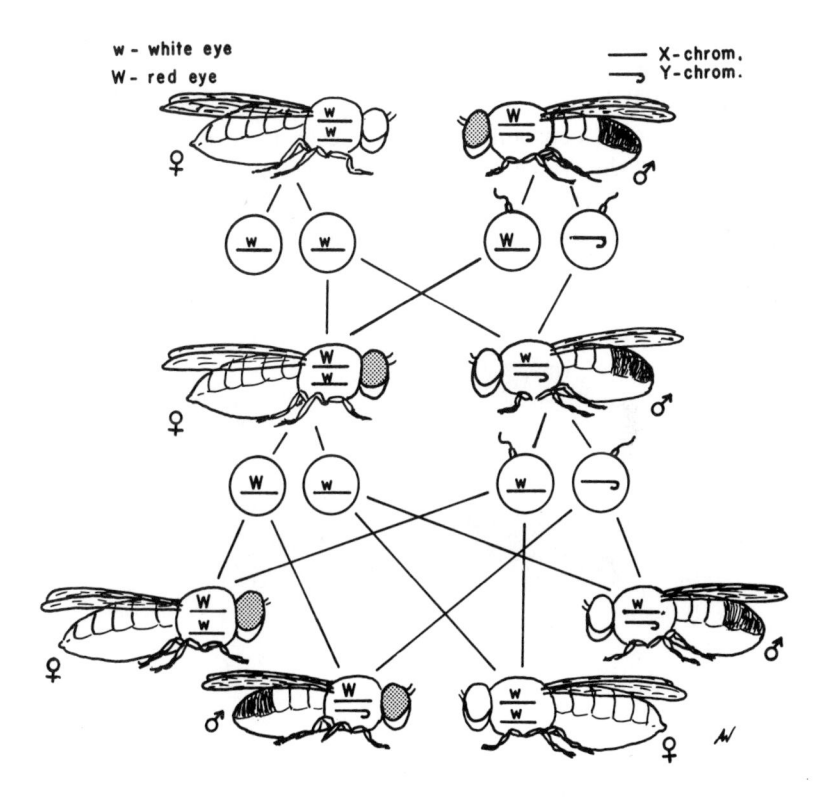

Fig. 8.2. *Diagram showing the transmission of sex-linked genes in Drosophila. Note that the male offspring receive their single X-chromosome from the female parent.*

parent, but the female receives an X-chromosome from each parent. If it is assumed that the gene for white eyes is recessive, then the females would be heterozygous and would show the dominant red. Males, on the other hand, would as a rule be haploid with respect to the genes on the X-chromosome because the Y-chromosome contains very few genes which are homologous to those on the X. Recessive genes will be expressed when there is no dominant allele present, so that the males will express these haploid genes. The term *hemizygous* is used to refer to such genes since they can be called neither homozygous nor heterozygous.

Hemophilia in Man. Hemophilia, bleeder's disease, is one of the most familiar afflictions associated with the human sex-linked genes. Even though it is comparatively rare, its effects are dramatic and its method of inheritance most unusual. The clotting time of the blood of persons who inherit this affliction is unusually long so that small injuries may bleed for many hours; thus these individuals are in constant danger of death from excessive bleeding. The Crown Prince of Spain, Prince Alfonso, had this disease and it brought about his death in Miami, Florida, as a result of rather minor injuries received in an automobile accident. This

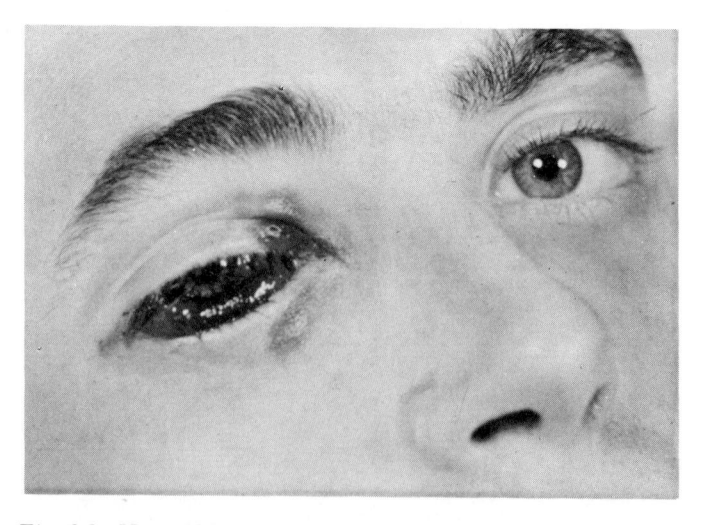

Fig. 8.3. Hemophilia in man, a sex-linked trait. The rupture of a small blood vessel under the skin of the eye lid resulted in extensive bleeding causing the condition shown here. (Courtesy J. V. Neel, Heredity Clinic, University of Michigan.)

death brought to mind other untimely deaths of members of the royal families of Europe as a result of hemophilia.

Apparently the gene arose as a mutation in a reproductive cell which produced Queen Victoria of England. She had one son with the affliction and two daughters who were carriers, a condition to be expected if she were heterozygous for the gene. The pedigree in Fig. 8.4. shows how the affliction spread to the different countries of Europe through intermarriage of the royal families. One of the most famous cases concerns the Tsarevitch Alexis of Russia. As a child he had repeated brushes with death because of excessive bleeding from minor injuries. The mad monk of Russia, Rasputin, obtained great power through the royal family because of his apparently successful treatment of the young Tsarevitch.

Hemophilia in Women. Recessive sex-linked genes, of course, can appear in females as well as males, but if such genes are rare the chance of two coming together from both parents is very small. For example, it is estimated that the frequency of hemophilic males at birth is about one in 10,000. There should be an equal number of heterozygous female carriers of the gene. If we assume that there is no lowered viability, then the chance of a marriage between a hemophilic man and a carrier female would be the square of $1/10,000$ or one in 100,000,000. About half of the female offspring would be homozygous and have hemophilia. This is one for each 200,000 female births. The estimated figure of 100,000 is much too large, however, because of the high death rate among the boys who have hemophilia—about three fourths die before adulthood. Also, because of their condition, many of the survivors might decide against marriage or having children.

In the above calculation, however, we assume random mating. If the gene is present in isolated regions where there is a high degree of intermarriage, then the chance of its becoming homozygous in a female will be correspondingly high. Such a region has been found in England; there, one family pedigree including 62 people had 13 males and 5 females with hemophilia.

It had been postulated that a female with hemophilia could not live beyond puberty because there would be excessive bleeding during her menstrual period. Actually, not only did these women survive beyond puberty, but four of them actually bore children. This may be explained in the light of some recent studies of hemophilia. Normal blood, when removed from the body and

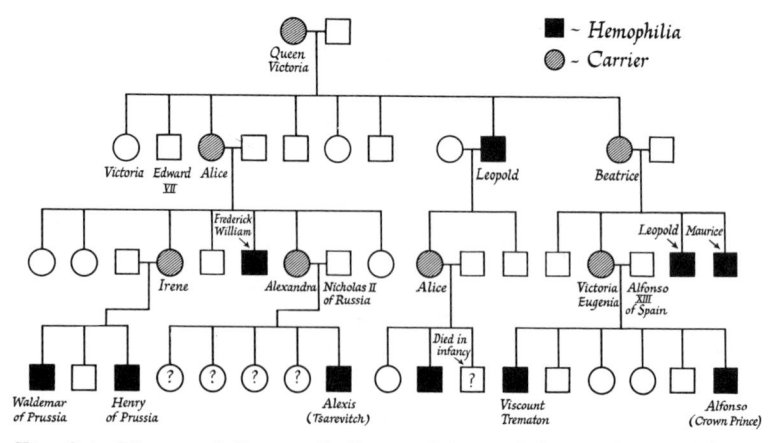

Fig. 8.4. The royal disease. Pedigree of hemophilia in the royal families of Europe. Apparently the gene arose as a mutation in the immediate ancestry of Queen Victoria of England and spread throughout the royal families with a great impact on history. (From Winchester, Genetics, Houghton Mifflin.)

placed in a container, will clot in about five minutes. In one pedigree of hemophilia it was found that the time required for clotting for the afflicted persons was as high as two hours. Another pedigree contained individuals that showed a clotting time ranging between 20 and 50 minutes, while a third pedigree revealed that the afflicted persons had a clotting time of only about 16 minutes. These results indicate that there may be several varieties of this gene (multiple alleles) which cause various degrees of severity of the disease. The cases in England could be explained by the fact that this is one of the alleles which reduces clotting time only slightly. Or, it could be a blood clotting disorder of a different kind, such as Christmas disease or parahemophilia, which is not due to the same gene.

Color Blindness. This condition in man is due to a sex-linked recessive gene and is much more common than the gene for hemophilia. One of the genes on the X-chromosome functions in the formation of the color sensitive bodies of the retina of the eye which are necessary for the distinction of red and green. There is a variant of this gene which makes such distinctions difficult. As in the case of the gene for hemophilia, there are varieties of this gene which cause variations in the degree of red-green color

SEX-LINKED GENES 105

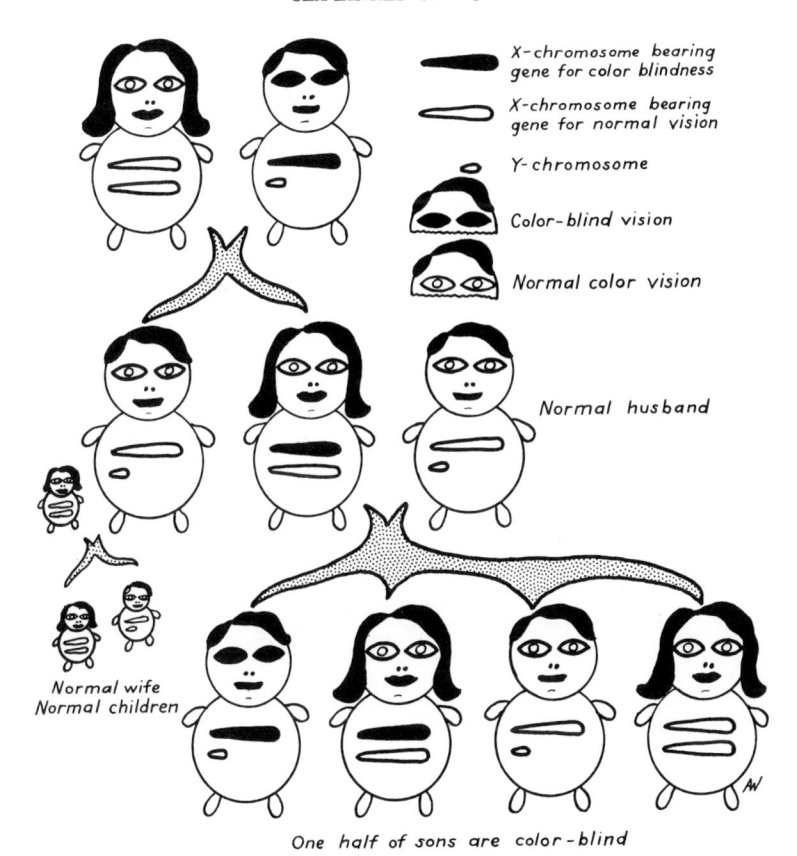

Fig. 8.5. Skip generation transmission of color blindness in man. (From Winchester, Biology, Van Nostrand.)

blindness. In its most extreme form there is complete inability to distinguish these two colors—red and green traffic lights have no meaning to persons expressing this gene. Other alleles range through several degrees to a mild form where the colors can be distinguished if they are brilliant and in good light, but are confused when the colors are of low intensity and the lighting conditions are poor.

As would be expected, the red-green color blindness is much more common among men than women. About 8 per cent of the men in the United States are color-blind when all degrees are included, but only 0.5 per cent of the women can be so classified.

Frequency of Expression of Sex-linked Genes according to Sex. Recessive sex-linked genes are expressed more frequently in the males where there exists the XY or XO method of sex determination. Since most deviations from the normal condition which are used as illustrations of sex-linkage are recessive, one may get the false impression that all sex-linked genes are expressed more frequently in males than in females. This is not true—dominant or intermediate genes will show more frequently in females since they have two X-chromosomes and, hence, twice the chance of having any particular sex-linked gene.

In *Drosophila*, for instance, there is a sex-linked gene for *bar eye*. Males hemizygous for this gene have the eye reduced to a narrow band. Heterozygous females have an eye in between the normal and the bar eye—it is called wide-bar. Homozygous females have the bar eye similar to that of the males. In any mixed population where the gene is present, we find that females show an effect of this gene about twice as frequently as males.

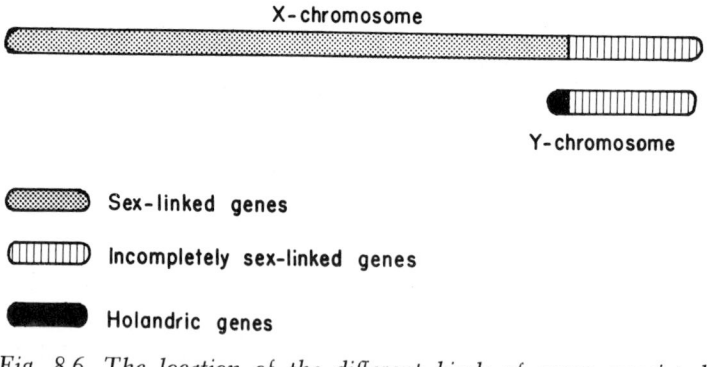

Fig. 8.6. *The location of the different kinds of genes associated with the sex chromosomes. The incompletely sex-linked genes are the only ones shared by both the X-chromosome and the Y-chromosome.*

In man there is a dominant sex-linked gene for defective enamel of the teeth. This allows excessive wear and the teeth become reduced to mere stubs in adulthood. About twice as many cases are found among women as among men.

Animals with Other Types of Sex Determination. In animals with the XO method of sex determination, there is no difference in the method of inheritance of the genes on the X-chromosome.

It follows the same pattern as the one described in the XY method.

In animals expressing the ZW method, the sex-linked genes lie on the Z-chromosome and the females are the hemizygous sex. The condition known as **barred feathers** (as found in Plymouth Rock chickens) may be used as a typical example of such sex-linkage. Barred is due to a dominant gene and non-barred to its recessive allele. Fig. 8.6. illustrates how this condition is inherited.

In animals with the honeybee method of sex determination, all genes are hemizygous in the males and all genes behave as if they were sex-linked. Drone bees receive no genes from a male parent—all of their chromosomes come from the queen, but the females express biparental inheritance.

INCOMPLETELY SEX-LINKED GENES

We have emphasized the fact that the Y-chromosome is made up largely of inert material and carries comparatively few genes, and that these few are transmitted in a certain manner. Some of the genes are allelic to some of the genes of the X-chromosome and, therefore, are diploid in the males as well as in the females. These genes are known as incompletely sex-linked genes. They behave very much as autosomal genes, but their association with sex-linked genes can be detected by studies of the pattern of transmission.

The gene for **bobbed bristles** in *Drosophila* was one of the first incompletely sex-linked genes to be discovered. Nine such incompletely sex-linked genes have been reported in man. These include: total color blindness, *epidermolysis bullosa* (eruption of skin blisters which may become malignant), *retinitis pigmentosa* (a dominant gene causing pigment on the retina and defective vision), *hemorrhagic diathesis* (a blood abnormality), and a form of *cerebral sclerosis* (a condition which causes mental retardation).

HOLANDRIC GENES

In most forms of life which employ the XY sex determination, there is a large part of the Y-chromosome with no counterpart on the X. Most of the chromosome appears to be devoid of genes, but any genes which might be present on this region would show a very unusual method of inheritance. Characteristics due to such genes would be transmitted from the male parent to all his male offspring and would show in every male offspring every generation. In *Drosophila* there are two genes which influence male fertility that fall into this category.

A human pedigree showing this type of inheritance was found in England during the eighteenth century. A boy born in 1717 developed a very thick skin covered with bristles, a condition known as *icthyosis hystrix gravior*. This inherited trait apparently arose as a mutation of a gene on the Y-chromosome. When the boy grew to maturity, he married and had six sons with the same condition, but all of his daughters were normal. During four more generations the characteristic was traced and found to appear in all the sons, but in none of the daughters. This appears to have been a case of holandric inheritance.

This type of inheritance has been suggested for 17 human genes, but a detailed study of the evidence shows that many of these could be cases of unusual distribution of traits due to dominant autosomal genes according to sex. Chance variation could account for them. In addition to the case mentioned above, cases which seem to have the most convincing evidence of being holandric are: *hypertrichosis* (long hair growth) of the ears and *keratoma dissipatum* (hardened lesions on the hands and feet). Even these, however, are somewhat doubtful.

SEX-LIMITED GENES

There are some genes that are expressed in only one sex. They have no particular relationship to the sex-chromosomes, and the great majority of them are autosomal, but there is something about the nature of sex which allows them to exert their action in only one of the two sexes.

In Cattle. This condition is understandable if we use milk production in cattle as an example. Just as the cow, the bull carries genes which influence this characteristic, but the bull obviously cannot express the genes he carries. He may, however, transmit genes for high milk production to his female offspring. Some bulls are so well endowed with such genes that they are known to breed calves which always yield greater milk than their mothers.

In Butterflies. In the clover butterfly the males are always yellow, but the females may be either yellow or white. A study of the inheritance of this characteristic shows that white is expressed in females when a certain dominant gene (W) is either heterozygous or homozygous—otherwise, they are yellow. Males, on the other hand, will be yellow even though they carry the gene W. Gynanders have been found which carry the gene for white and these are white on the female side and yellow on the male side.

In Birds. In vertebrate animals the sex-limited genes seem to be limited to one sex because of the presence of sex hormones. Nowhere in the animal kingdom is this more evident than in those birds whose sexual dimorphism is displayed most strikingly (as, for example, the peacock, the pheasant, etc.).

The pattern of feathering in the domestic fowl illustrates an interesting case of sex-limited genes. A recessive gene (h), when homozygous, causes the normal cock-feathering such as is found in the males of most breeds of chickens. The gene has no effect on the females, however, and they are all hen-feathered. There are some breeds of chickens, such as the Hamburgs, in which some of the males are hen-feathered. Breeding tests show that in this breed there are some genes for hen-feathering (H). When males receive one or two of these genes they will be hen-feathered. Still other breeds, such as the Sebright bantams, have hen-feathering in both sexes. These are all homozygous for the hen-feathered gene (HH). We can represent this relationship in the following way:

Genotype	Females	Males
HH	Hen-feathered	Hen-feathered
Hh	Hen-feathered	Hen-feathered
hh	Hen-feathered	Cock-feathered

We can demonstrate the effect of hormones on these gene expressions by experiments on breeds whose females are all hen-feathered and whose males are all cock-feathered. Castration of the male causes a bird to be produced which has the hen-feathering. Castration of the female and administration of the male hormone produces a cock-feathered chicken. Thus, it seems evident that the male hormone is necessary for the expression of the gene for cock-feathering.

In Man. Many sex-limited genes are known in man. The genes for a male beard, male voice, and male musculature normally are expressed only when the male hormone is present. The genes for feminine breast development, on the other hand, are expressed only when the female hormone is present.

SEX-INFLUENCED GENES

The genes which fall under this category may appear in both sexes, but they are so influenced by sex that they act as dominants in one sex and recessives in the other. A higher threshold of gene activity is required for their expression in one sex.

Coat Color in Cattle. The Ayrshire cattle are all spotted, but some have red spots on white while others have mahogany spots on white. The difference is due to a single gene. The mahogany and white is due to a gene which is dominant in males and recessive in females. For such genes it is customary to use the first letter of the characteristic which is dominant in males, so we use M to represent this gene. Homozygous MM cattle are mahogany and white regardless of sex, and homozygous mm are red and white regardless of sex. The heterozygous Mm, however, are mahogany and white if males, or red and white if females.

Horns in Sheep. A similar condition is found in sheep. The Dorset sheep have horns in both sexes, whereas the Suffolk have horns in neither sex. The Dorset are homozygous for the genes for horns (H) and the Suffolk are homozygous for the allele (h). When these two breeds are crossed we find that all the male offspring have horns and all the females are hornless. This shows that the gene for horns is dominant in the males, but recessive in females.

Baldness in Man. Baldness may arise from disease and other environmental causes, but most cases observed are due to heredity. The pattern of inheritance suggests a sex-influenced gene as the responsible agent. Men regularly transmit the characteristic to about one half of their sons, as would be expected if they are heterozygous for a dominant gene. Normal women with bald fathers sometimes have baldness in about one half of their sons as would be expected if the women are heterozygous. Women are bald much more rarely than men since they would have to be homozygous if the gene is sex-influenced. Bald women do occur, however, although it is easier for them to disguise the condition. Such bald women have all bald sons if their baldness is of the inherited type.

TYPICAL PROBLEMS AND ANSWERS

1. White eyes in *Drosophila* is due to a recessive sex-linked gene. Show the offspring to be expected from a cross of a white-eyed male and a homozygous normal female. Also, show the F_2 offspring.

Answer.

F_1: $wY \times WW = wW + WY$ (all have red eyes)
F_2: $wW \times WY = wW + wY + WW + WY$ (all females have red eyes; one half of males have red eyes and one half have white eyes).

2. Several cases of hemophilia in women have been reported in a pedigree from England. Show the expected children of such a woman married to a normal man.

Answer. $hh \times HY = hH + hY$ (all daughters normal; all sons have hemophilia).

3. A color-blind man marries a woman with normal vision who had a color-blind father. Show the expected children.

Answer. The woman must be heterozygous since she receives the single X-chromosome of her father and this must carry the gene for color blindness; thus:

$Cc \times cY = Cc + CY + cc + cY$ (one half of the sons and one half of the daughters are color-blind).

4. Optic atrophy (blindness due to atrophy of the optic nerve) occurs as a result of the action of a recessive sex-linked gene. A woman with optic atrophy marries a normal man. Their first child, a boy, has optic atrophy and is an albino. Show the ratio to be expected in future children.

Answer. Both parents are obviously heterozygous for the autosomal gene for albinism and the woman must be homozygous for optic atrophy. This is a dihybrid and with the short-cut method the results are:

3 normal pigmentation + 1 albino
× 1 male with optic atrophy + 1 female with normal vision
---
3 normal pigmentation males with optic atrophy + 1 albino male with optic atrophy + 3 normal females + 1 albino female.

5. In cats there is a gene for coat color that is both intermediate and sex-linked. One gene is for yellow and its allele is for black, but the heterozygote has a peculiar combination of yellow and black, a condition known as tortoise-shell. Show the types of offspring expected from a cross of a tortoise-shell female and a yellow male.

Answer. $I^bI^y \times I^yY = I^bI^y + I^bY + I^yI^y + I^yY$ (1 tortoise-shell female : 1 black male : 1 yellow female : 1 yellow male)

6. Barred feathers in the domestic fowl develop as a result of the action of a dominant sex-linked gene. Show the offspring to be expected from a cross between a barred hen and a non-barred rooster.

Answer. The female is the heterogametic sex in birds and will carry only a haploid set of sex-linked genes, hence:

$BZ \times bb = Bb + bZ$ (males are barred and females, non-barred).

7. Cock-feathering is an autosomal recessive in the domestic fowl and is sex-limited to the males. A hen has normal hen feathers, but is homozygous for cock-feathering. The hen is crossed with a rooster that has hen-feathering, but is heterozygous. Show the expected offspring.

Answer. h—cock-feathering in males H—hen-feather in males
 Females are hen-feathered regardless of these genes.

$hh \times Hh = hH + hh$ (one half of the males will be hen-feathered and one half will be cock-feathered; all females will be hen-feathered).

8. A mahogany and white cow is mated with a red and white bull. Show the ratio expected from this cross.

Answer. Both of these animals would be homozygous because the mahogany and white is recessive in the female and dominant in the male.

$MM \times mm = Mm$ (males are mahogany and white and females are red and white).

9. A man has many hard lesions on his hands and feet. He has five sons who have the same condition, but his three daughters are normal. One daughter is married and has two normal sons. One son is married and has a boy who has the same lesions. What type of inheritance would be suggested by this pedigree?

Answer. This would appear to be typical holandric inheritance since all sons receive their Y-chromosome from the father. The pedigree alone, of course, would not prove this inheritance—the case could be merely one of unusual chance distribution of the gene. We would need to include other pedigrees before drawing any definite conclusions.

10. Two hornless sheep are mated and about half of the male offspring from many crosses have horns while all the females are hornless. Show the genotype of the parents.

Answer. The gene for horns (H) is dominant in the males and recessive in the females. Since some of the male offspring have horns, the female parent must have been heterozygous. The genotype of the parents would be:

$$\text{Female: } Hh \qquad \text{Male: } hh$$

9: MULTIPLE ALLELES AND MULTIPLE GENES

Inherited characteristics which have been previously used for illustrations in our study have been for the most part those easily distinguished from one another. In man there have been mentioned such traits as albinism and normal pigmentation; in *Drosophila*, vestigial wings and wings of normal length; in garden peas, the phenomena of green peas and yellow peas. In some cases with intermediate inheritance, we have studied three classes—the red, roan, and white cattle. This relationship is not true of all characteristics which are inherited. Human stature is certainly affected by heredity, yet people do not fall into two or three distinct classes according to height. There is continuous variation from one extreme to another. The color of hair in man is another characteristic which has many varieties, and certainly this trait could not be explained on the basis of a single gene variation. We must assume that there are more than two varieties of genes in such characteristics. These genes may occupy the same locus of the chromosome, although one organism could carry no more than two. Such genes are known as multiple alleles. Or, the genes may occupy a number of different loci and are then known as multiple genes or polygenes.

MULTIPLE ALLELES

We could never discern the function of any particular gene if there were not some variant allele which causes a variation in phenotype. For example, we could not know that there is a particular gene in mice which contributes to normal hair growth if we had not found an allele that caused the hairless condition. Such variant alleles arise as mutations of existing genes. There is no reason to doubt the existence of more than one variety of alleles—it is logical to assume that there can be more than one possible variation of a gene. In the case of the sex-linked genes for both hemophilia and color blindness, we have learned that there are different alleles, known as multiple alleles, which cause a variety of degrees of these afflictions.

Drosophila Eye Color Alleles. The gene which produces *white eyes* in *Drosophila* was discovered in the genetics laboratory at Columbia University during the early years of genetic research (1910). The gene was a sex-linked recessive gene, and the normal wild-type allele contributed to the production of red eyes. Later, another sex-linked recessive gene was found which also affected eye color. This gene produced *eosin eyes*, a light orange-red color. When an eosin-eyed male was crossed to a white-eyed female the male offspring all had white eyes (they received their single X from the female parent) and the females had an eye color intermediate between eosin and white. Thus, it is evident that these two genes are alleles and both are allelic to the gene which contributes to the normal red eyes. This was the first case of multiple alleles discovered.

The discovery raised a problem about gene symbols. When white eyes was discovered it was given the symbol w, and W was used for the normal allele. As a symbol for eosin it was decided to retain the use of the same letter and to add the first letter of the new characteristic as a superscript, thus, w^e.

This was only the beginning of the discovery of alleles at this locus. Today fourteen different alleles are known. The wild-type W is dominant to all; the others, in general, show intermediate inheritance when crossed with one another. They are listed below somewhat in the order of intensity of the color of the eye.

W—red	w^b—buff
w^w—wine	w^t—tinged
w^{cc}—coral	w^h—honey
w^{bl}—blood	w^{ec}—ecru
w^c—cherry	w^p—pearl
w^a—apricot	w^i—ivory
w^e—eosin	w—white

Multiple Alleles in Other Forms. Many other cases of multiple alleles have been discovered. In rabbits there is a gene c which causes *albinism* and its wild-type allele C which functions in the reproduction of full color. A third gene was discovered which causes the rabbits to be white except for color on the ears, tail, feet, and nose region. This pattern is known as **Himalayan** and the gene symbol is c^h. Unlike the white-eye series in *Drosophila*, however, the gene for Himalayan was found to be dominant to the gene for albinism, but recessive to the wild-type allele C. A fourth allele was found that causes the fur to lack all yellow pigment; this results in a silver-grey appearance which is designated as *chinchilla* (c^{ch}). It was found to be dominant to the gene for Himalayan and also to the gene for albinism, but recessive to the gene for full color.

In man it has been found that the genes for the blood groups are due to multiple alleles. The genes for type A and type B blood are intermediate, but both are dominant to the gene for type O. (This characteristic and the genes for the Rh-factor are discussed in the next chapter.)

MULTIPLE GENES (POLYGENES)

Most characteristics do not result from the action of a single pair of genes, but from the combined action of a number of genes at different loci on the chromosomes. If we know of a variant of only one of these genes we may come to think of a particular characteristic as being produced by only one pair, but continued study often reveals variations at other loci which affect the same characteristic. We speak of the group of genes at different loci which are involved in the expression of one characteristic as multiple genes or polygenes. We can turn to our old stand-by, *Drosophila*, to illustrate this phenomenon.

Multiple Genes for Eye Color in Drosophila. The *white-eye* series of multiple alleles is located at 1.5 on the X-chromosome. It has been discovered that there is another gene which produces *scarlet eyes.* Genetic tests showed it to be located at 44.0 on the X-chromosome. Then, a third mutant was found that produced *purple eyes*—it was located at 54.5 on the second chromosome. This was rapidly followed by mutants at still other loci which caused eye colors of *carnation, cinnabar, brown, garnet, pink, ruby, vermillion,* and *sepia*. Thus, we see that the formation of pigment in the eyes is the result of the combined action of a group of genes at different loci—these are known as multiple genes.

When *Drosophila* are captured in the wild state, they almost

MULTIPLE GENES (POLYGENES)

always have the wild-type red eye. Hence, they would carry the normal alleles of all of these mutants mentioned above. When we find a fly with purple eyes, we know that it is homozygous for the gene for purple (pr/pr), but it also must carry at least one wild-type gene at each of the other loci which affect eye color.

In some cases, of course, a fly may be homozygous for more than one of the multiple genes involved in eye color. For example, a fly homozygous for the gene for vermillion eye color and the gene for brown eye color will have an eye which appears about the same as the white (this result is brought about by a single gene on the X-chromosome). The reason for this eye color is apparent when the gene action is studied. The gene for vermillion seems to prevent the formation of a brown pigment in the eye, but a bright

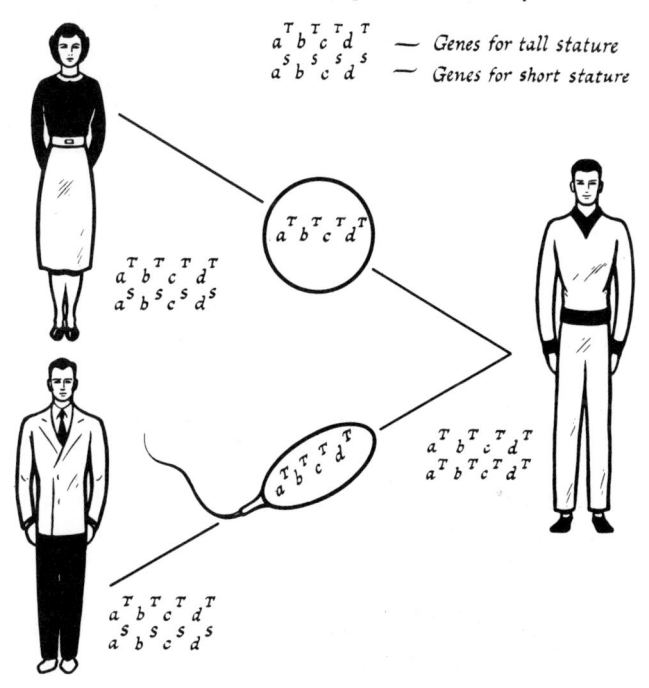

Fig. 9.1. Multiple gene inheritance in man. Body stature is influenced by a number of genes. The diagram above shows how parents of medium stature can have a child who is very tall because of the segregation of the genes for tall stature in the gametes of both parents. (From Winchester, Genetics, Houghton Mifflin.)

red develops and this is known as vermillion. The gene for brown prevents the vermillion pigment from developing and the eye is brown. When both genes interact, neither the brown nor the vermillion, the two major pigments of the eye, develops and the eye is white.

Multiple Genes for Coat Color. In other cases there may be an epistatic effect of one gene over the other, as was discussed in Chapter 5. In mice there is a gene for albinism (c) which is recessive, so that at least one (C) is necessary for the expression of any color of the coat. The particular color that is expressed, however, depends upon other genes. Most wild mice have the dominant gene (A) which causes the agouti pattern—a salt and pepper appearance caused by the fact that each hair has a yellow band near the tip. Its allele (a) causes each hair to be uniformly colored without the band. At still a third locus there may be a gene (B) which causes a black pigment in the hairs, and it has an allele (b) which forms a brown pigment. A gene at a fourth locus (D) allows full intensity of the color to develop, but its allele (d) causes a paler (dilute) shade of the color. If the mouse is homozygous for the first gene mentioned (c/c), then it makes no difference what the others are—the mouse is an albino nonetheless.

From the foregoing we may perceive how complex the interactions of genes can be as inherited characteristics are developed. A fuller discussion of this topic is found in Chapter 15.

MULTIPLE GENES AND QUANTITATIVE CHARACTERISTICS

Great quantitative variation is possible when one characteristic is influenced by multiple genes so that it may be expressed in different degrees. Examples from several forms of life include: the color of wheat, size in children, and skin color in man.

Color in Wheat. The great Swedish geneticist, Nilsson-Ehle, obtained a number of varieties of wheat with kernels of different color. A cross between wheat with dark red kernels and wheat with white kernels gave offspring with an intermediate shade of red kernels. An *inter se* cross of this F_1 yielded five varieties of offspring with the color ranging from deep red to white. The ratio was 1:4:6:4:1. Such results would be expected if there were two gene loci involved in pigment production. We can assume that genes at these loci may be *contributing* (pigment producing) or *neutral*. Four contributing genes for the red pigment would yield the darkest red and the others would show red in proportion to the number of contributing genes present. The following table shows the expected F_2.

No. Neutral Genes	No. Contributing Genes	Color	Proportion
4	0	white	1/16
3	1	light red	4/16
2	2	medium red	6/16
1	3	dark red	4/16
0	4	darkest red	1/16

Size in Chickens. The effect of a greater number of multiple genes may be illustrated by a cross involving chickens of different sizes. Punnett crossed the small **Sebright Bantam** with the large **Golden Hamburg** and obtained birds of an intermediate size in the F_1. In the F_2 there existed a variation, as would be expected, but to his surprise there were a few chickens that were larger than either of the P_1 and some that were smaller than either. Where could the genes come from to cause these extremes of size?

On the basis of the results obtained, Punnett reasoned that there must be four pairs of genes involved in size in this particular cross. We might assume that the Hamburgs are homozygous for three pairs of *contributing genes* for large size and homozygous for one pair of *neutral genes* which would tend to cause small size. If we allow capital letters to represent the genes for large size without indicating dominance, the genotype of the Hamburgs would be: AABBCCdd. The Bantams, on the other hand, could be homozygous for D, but all the other genes would be for small size. Their genotype would be read as *aabbccDD*. The F_1 would be heterozygous for all four genes and would be intermediate in size, but in the F_2 there would be a few chickens that would receive the genotype AABBCCDD and these would be larger than the Hamburgs. Conversely, some could be *aabbccdd* and be smaller than the Bantam.

This principle is used to provide new genes for selection in plant and animal breeding. Sometimes a rather desirable variety is crossed with a poor variety and then, through selection, it is possible to pick up some good characteristics from the poor variety and make the product better than the original. The yield of corn has actually been increased by crossing a high yielding variety with one that has low yield and then selecting for high yield. Even the poorest of varieties seem to have some genes which can improve those that are the best.

Skin Color in Man. The amount of melanin pigment deposited in the skin of man varies according to the amount of exposure to sunlight, but even without such exposure there exists a great variation among the different peoples of the earth. The variation in pigment is continuous and thus suggests multiple gene inheritance.

We have an opportunity to study the method of inheritance of this trait in interracial marriages between members of the Negro and Caucasian races. Such marriages produce children with an intermediate shade of skin, known as mulatto. Studies of marriages between two mulattoes who are children of interracial marriages show that the offspring may vary all the way from the fair skin of the Caucasian to the more heavily pigmented skin of the Negro. An analysis of the results of such marriages by Curt Stern of the University of California indicated that there must be at least four contributing genes for pigment deposition involved in the gene differences between the two races. This would give an expectation of one white child out of every 256 from mulatto marriages, and a like number with the full Negro pigmentation. The variation of skin pigmentation within the Caucasian race shows that certainly there must be other genes involved, but this study concentrates only on those genes which bring about the skin color distinctions between the two races.

Estimating the Number of Multiple Genes. It is possible to arrive at an estimate of the number of multiple genes involved in the production of any characteristic if one can obtain sufficient F_2 results to find the ratio of individuals that show one extreme or the other. Thus, in wheat, where two gene loci were involved in producing the color of the kernels, it was noted that $1/16$ of the kernels were white and the same number were of the darkest red color. As can be seen in the following table, the fraction rapidly becomes smaller as the number of genes involved increases.

Number of Pairs of Genes Involved	Fraction of Offspring That Show Most Extreme Variation
1	1/4
2	1/16
3	1/64
4	1/256
5	1/1024
6	1/4096

USE OF THE STANDARD DEVIATION

In evaluating the degree of variation brought about by the quantitative action of multiple genes, the standard deviation is frequently used as an objective measure. This measure is a special form of standard error which shows the degree of variation from the mean. It is obtained by taking the square root of the average squared deviations from the mean and is represented in mathematical symbolism as:

$$\sigma \text{ (standard deviation)} = \sqrt{\frac{\Sigma f d^2}{n}}$$

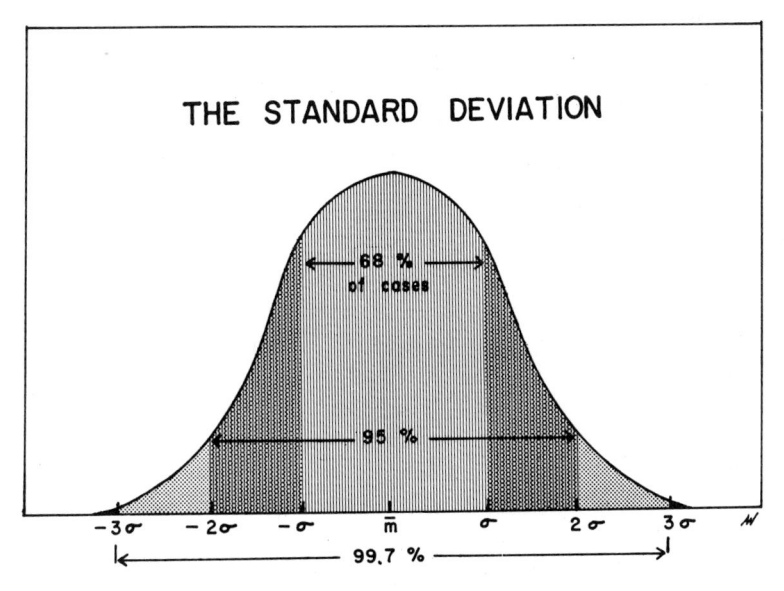

Fig. 9.2. *The curve of normal distribution. The standard deviation is valuable in determining the percentages that lie within specific degrees of variation from the mean.*

We can illustrate the use of the standard deviation measure in a study of height in man and in a comparison of two groups as to relative variability.

Height in Man. If we should choose a large group of men as, for example, students at a military school, and plot their heights on a curve, we will almost certainly find a typical bell-shaped curve of normal distribution. The largest number of men will be grouped around the mean with the numbers gradually getting smaller as we approach the extremes. Fig. 9.2. shows such a curve. Suppose we find that the mean is sixty-eight inches. When we calculate the standard deviation we find that it is three inches. In other words, 68 per cent (about two-thirds) of the men will vary no more than three inches from the mean. About 95 per cent will vary no more than twice the standard deviation (six inches from the mean) and only 0.3 per cent will lie outside the area which varies three times the standard deviation.

Of course, we realize that environment has a great bearing on human stature, but the greater part of this variation can probably be assigned to multiple genes.

Comparison of Two Groups as to Relative Variability. The standard deviation is valuable in demonstrating the relative degree of homozygosity in different populations. Quantitative characteristics which are influenced by heredity would naturally be expected to show a greater degree of variation in a heterogeneous mixed population than would be the case in a population which was rather closely inbred.

We can illustrate this point with a study of the weight of chickens after eight weeks of special, uniform feeding. This is the age when they can be marketed as fryers and it is important for poultrymen to have a product which is rather uniform in size. The first table shows how the standard deviation is calculated for one group of chickens hatched from the general barnyard run of eggs. The second table shows the same calculation for a breed which has been closely inbred for many generations. We can see that, although the mean for the two groups is almost the same, the standard deviations are quite different. The inbred group would be homozygous for more genes which influence body size. Thus, the standard deviation yields an objective way of indicating the degree of variation from the mean and is especially useful in comparing two groups for degree of variation.

Size Distribution of Chickens in Breed A

Wt. range in pounds	Mid-class value v	Number of chickens in each group f	fv	Deviation of class value (v) from mean d	d^2	fd^2
2.5–2.7	2.6	8	20.8	−.4	.16	1.28
2.7–2.9	2.8	22	61.6	−.2	.04	.88
2.9–3.1	3.0	40	120.0	0.0	.00	0.00
3.1–3.3	3.2	18	57.6	.2	.04	.72
3.3–3.5	3.4	12	40.8	.4	.16	1.92
Σ		100	300.8			4.80

$$\overline{m} \text{ (mean)} = \frac{\Sigma fv}{n} = \frac{300.8}{100} = 3.008$$

$$\sigma \text{ (stand. dev.)} = \sqrt{\frac{\Sigma fd^2}{n}} = \sqrt{\frac{4.90}{100}} = .219$$

Size Distribution of Chickens in Breed B

Wt. range in pounds	Mid-class value v	Number of chickens in each group f	fv	Deviation of class value (v) from mean d	d^2	fd^2
2.5–2.7	2.6	4	10.4	−.4	.16	.64
2.7–2.9	2.8	16	44.8	−.2	.04	.64
2.9–3.1	3.0	60	180.0	0.0	.00	.00
3.1–3.3	3.2	14	44.8	.2	.04	.56
3.3–3.5	3.4	6	20.4	.4	.16	.96
Σ		100	300.4			2.80

$$\overline{m} = \frac{\Sigma fv}{n} = \frac{300.4}{100} = 3.004$$

$$\sigma = \sqrt{\frac{\Sigma fd^2}{n}} = \sqrt{\frac{2.80}{100}} = .1673$$

TYPICAL PROBLEMS AND ANSWERS

1. In *Drosophila* the eye color genes for white, apricot, and carnation are all sex-linked recessives. How would you determine whether these genes are multiple alleles or multiple genes?

Answer. Cross each gene with the remaining other two and study the phenotypes of the females. (Since these genes are sex-linked, only the females will be heterozygous, so we ignore the hemizygous males.) If a cross yields a fly with the wild-type red eye, you will know that the two genes are not alleles since each recessive gene will be suppressed by a dominant allele. On the other hand, if the result were females with eyes like one of the parents or somewhat intermediate between the two parents, this would indicate that the two genes are alleles.

2. Results obtained from the crosses described in Problem 1 were:

PARENTS		OFFSPRING	
Female	Male	Female	Male
white ×	apricot	light apricot	white
white ×	carnation	red	white
apricot ×	carnation	red	apricot

From these results determine the relationship of the genes.

Answer. White and apricot are alleles, but carnation lies at a separate locus.

3. A rabbit with a chinchilla coat is crossed with a rabbit with a fully colored coat. Would it be possible for both albino and Himalayan coats to appear in the offspring? Could either appear?

Answer. No, they could not express both, but they could have one coat or the other. The gene for full color and the gene for chinchilla are both dominant to the genes for albinism and for Himalayan coat; hence both parents could carry the gene for albinism and thus have an albino offspring. The same is true for the Himalayan coat, but if one parent carried the gene for albinism and the other carried the gene for Himalayan coat, then neither of these genes could become homozygous in the offspring.

4. In tomatoes assume that there are three multiple genes which are involved in the size of the fruit. Homozygous *abc/abc* are the smallest and have an average fruit weight of four ounces under good growing conditions. ABC are contributing genes, each of which adds

an average of one half ounce to the weight. Homozygous ABC/ABC have an average weight, therefore, of seven ounces. Show the average weight of the offspring of a cross between a plant bearing the smallest fruit with one bearing the largest.

Answer.

ABC/ABC × abc/abc = ABC/abc   Fruit is 5½ oz. (4 + 1½)

5. A 5 oz. tomato is crossed with a 5½ oz. tomato, and of a number of plants produced there is one that has fruit averaging 6½ oz. Show the possible genotypes involved.

Answer.

5 oz. parent could be *ABc/abc*   5½ oz. parent could be *ABC/abc*

6½ oz. offspring would receive *ABc/ABC*

(The 5 oz. parent could have been homozygous for the *a* or *b* as well as the *c*.)

6. In rare cases a child of extremely high intelligence will be born to a couple who are of only average intelligence. Explain how this might come about on the basis of multiple gene inheritance of intelligence.

Answer. Both parents evidently carried some genes for high intelligence which act as contributing genes. Through chance segregation of the genes, this child happened to receive all or most of the contributing genes for intelligence from both parents and developed a potential brain capacity much greater than that of either parent.

10: GENETICS OF THE HUMAN BLOOD GROUPS

The human blood shows many inherited variations. The variations are so extensive that a person's blood is almost as distinctive as his fingerprints as a means of identification. The variations are inherited in such a clear-cut manner that they may be used as a basis for the establishment of true relationships in cases of disputed parentage. The blood groups are also important in blood transfusion and are a factor in some cases of infant deaths. The genes responsible for the varied blood characteristics give us some very good examples of multiple alleles and multiple genes in man.

THE ABO BLOOD GROUPS

The existence of the blood groups was not known until the beginning of the present century, but the incompatibility of certain bloods when mixed was recognized as early as the eighteenth century. Blood transfusions were tried on men who had bled excessively and in some cases they were successful, but in others the recipient died. As a result of experimentation with blood that was mixed outside the body it was found that sometimes a smooth mixture resulted, but in other cases the blood cells would adhere in clumps somewhat resembling milk when curdled with acid.

Discovery of the Blood Groups. The explanation for this strange reaction became better understood about 1900 when Karl Landsteiner made a revealing study on human blood. He separated blood cells from the plasma (serum) and sometimes recombined these elements. He noted that a smooth combination always resulted when the cells were recombined with the plasma from which they had been removed, but when plasma from one person was mixed with cells from another the mixture would be smooth in some cases but the cells would clump (agglutinate) in other cases. He found that people could be classified into four distinct groups on the basis of these studies. These groups were first designated by Roman numerals (I, II, III, IV), but as they became better understood they were classified as O, A, B, and AB.

Blood Antigens and Antibodies. The agglutination reaction of the blood can be traced to the presence of antigens in the red blood cells and corresponding antibodies in the plasma. Antigens are substances which are a part of protein matter which, when introduced into the body of a higher animal, are capable of stimulating the production of specific antibodies. Generally, this reac-

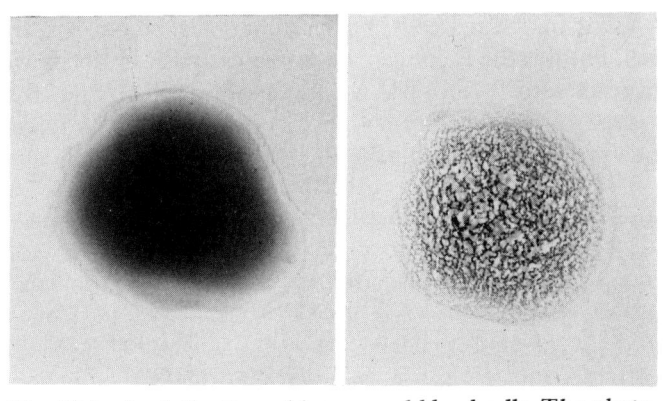

Fig. 10.1. Agglutination of human red blood cells. The photograph on the right shows how the cells clump together when a drop of blood from a person who is type A is mixed with serum taken from a person of type B. On the left it can be seen that there is a smooth mixing without any agglutination when blood from a person with type A is mixed with serum from another person with type A.

tion has a protective function. When disease germs invade the body, their antigens stimulate the production of antibodies which react against the foreign antigens. The reaction may cause the invading germs to be clumped or otherwise be so affected that they can be overcome by the body. These antibodies are highly specific and will react with one antigen only. For example, antibodies which are generated as a result of the contact with smallpox antigen will not be effective in protecting a person against polio. Likewise, a person may have hay fever attacks when he inhales large quantities of pollen of the short ragweed because he has developed antibodies against the antigen of this ragweed pollen. These antibodies, however, will not cause trouble when pollen from another plant is inhaled, even though it is a ragweed of a different species, as, for example, the giant ragweed.

The A and B Antigens and Antibodies. Landsteiner found two distinct antigens in the red blood cells of man and designated them as A and B. He found that the cells could contain either antigen or both antigens, or they could contain neither. This situation makes possible the four combinations of the four blood types. Strange to say, however, he also found that the blood serum always contained antibodies which would react with the antigens not present in the red cells. For example, if a person had the A antigen, but not the B antigen, he would carry the antibody which would react with B (anti-B). A person with neither A nor B antigens (type O) would carry both anti-A and anti-B. The following table shows this relationship.

Blood Type	Antigens Present in Red Blood Cells	Antibodies Present in Serum	Approximate Frequency in U. S. (in percentage)
O	neither	anti-A & anti-B	47
A	A	anti-B	40
B	B	anti-A	10
AB	AB	neither	3

Importance in Blood Transfusions. With this knowledge of A and B antigens and antibodies, we can understand why one person's blood will agglutinate in some cases when mixed with blood from another person. A transfusion can be given from a donor to a recipient of the same blood type without difficulties because there will be no antibodies present in the plasma of the recipient which would cause agglutination of the cells being transfused. If type A blood were given to a type O person, however, the anti-A in the plasma of the recipient would clump the type A cells and death would most likely result. It is always desirable to use blood of the same type in transfusion, but in an emergency it is possible to use a different type provided that the cells being given will not be clumped by the plasma in the recipient's body. Thus type O can be given to type A because the anti-A present in the plasma of type O blood will be rather quickly diluted by the plasma in the recipient's body; agglutination of the cells will not occur unless the antibodies have a certain concentration (titre). This fact makes it possible to give plasma transfusions without typing—the antibodies which may be present in plasma will not reach a sufficient titre to cause clumping of the cells in the body.

There is one question which has not yet been satisfactorily answered. Why are the anti-A and anti-B antibodies present in the human blood when there has been no apparent contact with the corresponding antigens? So far as is known, these are the only antibodies which develop in this way. A person with type B blood will always have anti-A in his plasma even though he has never contacted the A antigen, at least so far as science has recorded. There will be other antibodies in the plasma, but all of these can be traced back to contact with the specific antigen that stimulated the antibody production. We sometimes speak of the antibodies produced in response to contact with an antigen as immune antibodies, while those without such contact are called the normal antibodies.

Method of Typing Blood. Blood typing may be done by using comparatively simple techniques. Only two types of sera are needed—one from a person with type A blood and one from a person with type B blood. The former will contain anti-B and the latter will contain anti-A. The reaction of each of the four types of blood with these sera is shown in the following table.

Blood Type	Reaction with Anti-A (from Type B Blood)	Reaction with Anti-B (from Type A Blood)
O	no agglutination	no agglutination
A	agglutination	no agglutination
B	no agglutination	agglutination
AB	agglutination	agglutination

The cells agglutinate in large clumps when the blood is mixed with the serum; thus, on a slide they can be seen plainly with the naked eye.

THE BLOOD TYPES

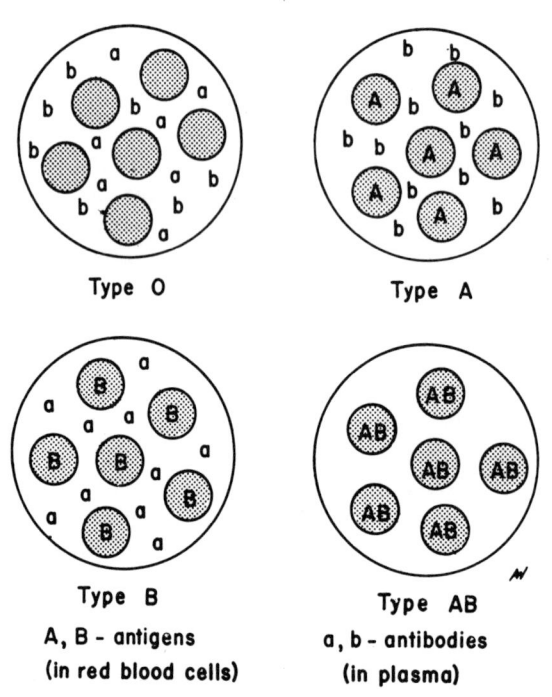

Fig. 10.2. The antigen and antibody constitution of the four blood types.

Type O

Type A

Type B

Type AB

A, B - antigens
(in red blood cells)

a, b - antibodies
(in plasma)

Inheritance of the ABO Blood Groups. Multiple alleles at an autosomal locus account for the inheritance of the ABO blood groups. There is one gene for the production of the A antigen. It appears as a dominant since it will be expressed whenever it is carried, either in a single or a double dose. The gene for the B antigen is the same, but a person heterozygous for these two genes develops both antigens and thus has type AB blood. This condition is similar to intermediate inheritance, but since there is no apparent lessening of the antigen in the heterozygote, it might be more properly called a lack of dominance of one gene over the other. Each gene acts independently and is not suppressed by the presence of the other. The third gene in the allelic series produces neither A nor B antigens. Hence, a person homozygous for this gene will be type O. This gene appears to be recessive to the other two, for a person will not be type O when heterozygous for either of the other genes.

As a basic letter symbol for these genes it was decided to use I (Isohemoglutinogen) and the three alleles are I^A, I^B, I^O. The possible genotypes of the four blood types are:

Blood Type	Possible Genotypes
O	$I^O I^O$
A	$I^A I^A$ or $I^A I^O$
B	$I^B I^B$ or $I^B I^O$
AB	$I^A I^B$

MEDICO-LEGAL APPLICATIONS OF BLOOD GROUP INHERITANCE

The clear-cut nature of the inheritance of the blood groups makes very valuable evidence possible in cases where there is any question of true parentage. Such cases as those involving mixed babies in the hospital or paternity suits are typical of the cases which may be solved by the use of scientific knowledge in this field.

In California there was a typical case of a paternity suit in which blood typing was used as evidence. A male movie star was being sued for the support of a child born to a lady acquaintance. Blood typing showed that he could not have been the father—the child was B, the mother was A, and the accused man was O. The true father must have been either type B or AB. The following table shows the different types of children which can result from different combinations of parents.

Blood Type of Parents	Possible Blood Types of Children
O × O	O
O × A	O, A
O × B	O, B
O × AB	A, B
A × A	A, O
A × B	O, A, B, AB
A × AB	A, B, AB
B × B	B, O
B × AB	A, B, AB
AB × AB	A, B, AB

The table shows that there will be many cases where the results of blood tests cannot exclude a person who may be falsely accused in a paternity suit, if this is the only information available. Had the movie star been type B he would have been a possible father of the child, but since there are many other men who carry the B antigen even this fact would have offered no conclusive evidence that he was the father of the child. Fortunately, from this standpoint, it has become possible to recognize three different varieties of the A antigen by special clinical tests. This increases the recognizable type to eight—three of A, three of AB, and one each of O and B.

A still further breakdown has recently become possible with the discovery that the gene for O produces a weak antigen of its own. It has no clinical significance because human beings do not produce any antibodies against it except very weakly in rare cases. When type O cells are injected into certain animals, however, a serum can be obtained which will react with the O antigen and cause agglutination. This makes it possible to distinguish heterozygous type A and type B blood from the homozygotes. A type A person with the genotype $I^A I^O$ would have blood that would show agglutination with this serum containing the anti-O, but a homozygous type A person would not react with the anti-O.

THE SECRETOR TRAIT

Continued study of the blood types has shown that some people have the blood antigens in the various body secretions, such as those from the eyes, salivary glands, and mammary glands. It is thus possible to identify blood types from a bit of dried saliva that may have been used in moistening an envelope or a bit that may have remained on a discarded cigarette butt. When serum with anti-A is mixed with the secretions from a secretor with type A blood, there will be an antigen-antibody reaction, but it cannot be seen. There are no cells which will be clumped, but the anti-A will be neutralized in the reaction. Later, if type A blood cells are added they will not clump. This reaction is one method of testing for the secretor trait.

Studies investigating why some people are secretors and some are not show that the secretors have water soluble antigens which can pass out of the red cells and thus be present in the body secretions. Those who are non-secretors do not have these water soluble antigens; hence such antigens do not pass out into the secretions. The secretor trait is inherited as a dominant (S), and the recessive allele (s) causes the non-secretor trait when homozygous. The ratio of secretors to non-secretors in the United States is about 77%:23%.

THE M AND N BLOOD ANTIGENS

As research progressed on blood antigens, it occurred to some investigators that the human blood cells might contain antigens for which the human body does not produce antibodies. (The anti-O has already been mentioned in this connection.) In 1927 the scientists Landsteiner and Levine injected human blood into rabbits and found that immune sera were produced which would react with some human blood samples but not with others. Further investigation showed that this could be explained on the basis of two antigens produced by two allelic genes. One was designated as M and the other as N and the corresponding genes as M and M^N. A person heterozygous for both genes would produce both antigens, as was the case with the A and B antigens. So far, however, no counterpart of O has been found. All persons produce either one antigen or the other, or both.

As was true with anti-O, these antibodies have no clinical significance because they are not produced by human beings—at least not in sufficient quantities to cause any difficulties in transfusion. The discovery was important, however, because it extends the known varieties of blood antigens and has valuable theoretical and medico-legal applications. In the population of the United States the proportions of the MN series are: 29%M:50% MN:21% N.

In 1947 a variety of the M antigen (M^s) was discovered, and in 1951 a similar variety of the N antigen (N^s) was found. These discoveries extended even further the possible varieties of human blood which could be distinguished by proper tests.

THE RH BLOOD ANTIGENS

Landsteiner and Wiener and other researchers found another interesting series of antigens which became known as the Rh antigens. These were first discovered when red blood cells from a rhesus monkey were injected into rabbits. After allowing time for sensitization, some serum was removed from the rabbits and mixed with the blood of the monkey. It was found that cell agglutination was caused. When human blood was mixed with this immune serum it was found that the cells would agglutinate in some cases and remain unclumped in others. Thus, it appeared that some people had an antigen in common with the rhesus monkey while others lacked this antigen. It was called the Rh factor from the first two letters of the type of monkey used. About 85 per cent of the people in the United States were found to have the antigen and were thus designated as Rh-positive. The remaining 15 per cent of the population were termed Rh-negative. The gene for the antigen appeared to be inherited as a simple dominant over the gene which did not produce the antigen.

Negative mother
Negative child
Baby normal

Same mother
Positive child
Baby normal
Mother sensitized

Same mother
Second positive child
Baby born with
erythroblastosis

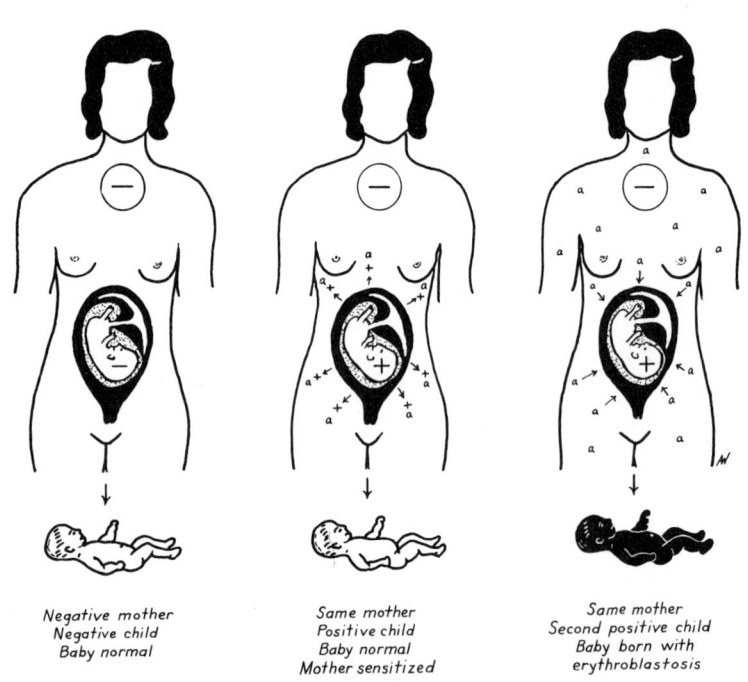

Fig. 10.3. *The Rh factor as an agent in causing erythroblastosis in babies. When a negative woman marries a positive man she may become sensitized when she bears a positive child and future positive children may be injured by the antigen-antibody reaction.* (From Winchester, Biology, Van Nostrand.)

It was found that the Rh antigen differed from the A and B antigens in that there were no normal antibodies for it in the serum of Rh-negative persons. The Rh differed from the M and N antigens in that the human body could produce the antibodies, and that could be clinically significant.

The Rh Antigens and Blood Transfusions. If an Rh-negative person receives a transfusion of blood from an Rh-positive person, the presence of this foreign antigen may stimulate the production of anti-Rh. The reaction will not affect the cells received during the transfusion because they will be worn out and replaced before the titre of antibodies is sufficiently high to cause agglutination. Should a second transfusion be given at a later time, however, there will be agglutination and death may result. Hence, it is undesirable to give positive blood to a negative person at any time. Today blood is usually typed for the Rh factor as well as for the ABO series.

Maternal-fetal Incompatability. The Rh factor also has great significance with regard to possible complications during pregnancy. Almost every person reading these lines knows of cases where a child has died at or shortly after birth because of this incompatibility. The difficulty may arise when a woman who is Rh-negative has a high titre of anti-Rh in her plasma and bears a child who is Rh-positive. The combination can happen only if her husband is Rh-positive, since the gene for the Rh factor is dominant. Also, the mother must have been previously sensitized by contact with the Rh antigen. This can come about by transfusion of Rh-positive blood or it might arise as a result of carrying an Rh-positive child.

Normally, there is no interchange of blood between a pregnant woman and the fetus within her body, but during the latter part of the pregnancy it is possible that the weight and movements of the embryo may cause some capillary breakage in the placenta and there could be some seepage of blood into the woman's body. If the woman is negative and this blood is positive, a sensitization can result. The first child will not be affected because of the time involved in building the antibodies. When a second positive child is conceived, however, the titre of antibodies may be sufficiently high to affect the blood cells of the embryo. The second child may be born with *erythroblastosis fetalis*, which is characterized by (1) severe anemia due to the hemolysis of the red blood cells and

(2) consequent jaundice as the blood vessels in the liver become clogged with broken cells and bile is absorbed by the blood. Also, the blood contains immature, nucleated red cells which are not efficient carriers of oxygen.

In many cases a negative woman can bear a positive child without becoming sufficiently sensitized to affect subsequent positive children. There is a real danger that she will, however, and a knowledge of the possibility can alert her physician to take precautionary steps to save the child in case it is born with erythroblastosis. Extensive blood transfusions can replace much of the damaged blood and save the child during the critical days immediately after birth.

Even though a man is positive and his wife is negative, it is possible that they will have some negative children if he happens to be heterozygous. This, coupled with the fact that the wife may bear some positive children without being sensitized lessens the danger. Also, as we shall soon learn, there are varieties of the Rh antigen and only one of these varieties is likely to cause antibody production. About 12 per cent of all marriages in the United States are of a positive husband and a negative wife. A study in England, where the percentage is about the same, showed that one child in each 23 born to negative women with positive husbands had the hemolytic erythroblastosis. Since a good proportion of these are first-born and some of the other children are negative, we can see that, while the over-all chance is rather low, there is a definite danger to positive children after the first child.

Complexities of the Rh Antigen Inheritance. Extensive study has been made on the Rh factor, and its complexities have been thus revealed. Just after subdivisions of the A antigen had been found, three major subdivisions of the Rh antigen were discovered. A. S. Wiener proposed the theory that there are eight multiple alleles at the locus for the Rh antigen. This number was considered possible because, according to his theory, some of the alleles produced two or three of the positive antigens. This is shown on page 137.

It was found that the human body produced antibodies readily against the R_o antigen, but rarely against the others. Hence, a negative woman had little to fear if her positive husband did not include a gene which produced the R_o antigen in his genotype. Erythroblastosis fetalis is caused by this antigen in 90 per cent

Gene Symbol	Positive Antigens Produced
r	none
R_o	R_0
R'	R'
R''	R''
R_1	R_o & R'
R_2	R_o & R''
R_x	R_o R' & R''
R_y	R' & R''

of the cases. Unfortunately, however, this is by far the most common positive antigen.

As if this did not complicate the picture enough, the geneticist Phillip Levine found that the blood cells of negative persons could induce the formation of antibodies, just as the blood cells of type O persons could produce antibodies. There existed three varieties of the antibodies, designated as Hr antigens, which corresponded to the three varieties of the Rh antigens. Thus, the number of possible alleles was doubled.

The English geneticist R. A. Fisher proposed a theory which was somewhat simpler to understand and which has found wide acceptance. He suggested that there are actually three genes at different loci, but lying very close together, which account for the antigens. These genes are known as pseudoalleles—that is, they act like alleles, but are actually multiple genes lying side by side and affecting the same characteristic. According to this theory, a person carries six genes for the Rh complex, three on each of the two homologous chromosomes involved. Fisher used the letters CDE to represent the three genes, with the capital letters standing for the positive antigens and the small letters for the negative condition and the negative antigens. Thus, a person would be Rh-positive if he carried even one of the capital letters—he might be cDe/cde. The gene D produces the antigen designated as R_o by Wiener; C is equivalent to R', and E to R''. A person's genotype can be determined by the reaction to the six anti-sera: anti-D, anti-C, anti-E, anti-d, anti-c, anti-e. The frequency of the genotypes most often encountered is given in the table on the following page. The table reveals that there are still other combinations of genotypes which would be possible, but those not shown are so rare that they have little likelihood of appearing in routine studies.

Genotypes		Approximate Frequency in Population in Percentage
Wiener's Notation Method	Fisher's Notation Method	
r/r (neg.)	cde/cde	15.00
R_1/r	CDe/cde	35.00
R_1/R_1	CDe/CDe	20.00
R_2/r	cDE/cde	12.00
R_2/R_2	cDE/cDE	2.00
R_1/R_2	cDE/CDe	13.00
R_o/r	cDe/cde	2.00
R'/r	Cde/cde	0.75
R''/r	cdE/cde	0.85
R_o/R_o	cDe/cDe	rare
R'/R'	Cde/Cde	rare
R_x/r	CDE/cde	rare
R_y/r	CdE/cde	rare

This is as far as space will permit us to go into this subject, but there is more to the story. For example, it has been possible to obtain subgroups of the C and D antigens and there is good evidence for an F antigen according to Fisher's theory.

OTHER BLOOD ANTIGENS

Still other blood antigens have been discovered and, no doubt, will continue to be discovered in future research. As an example, a Mrs. Kidd bore a child who showed some signs of antigen-antibody hemolysis, yet there was no incompatibility for any of the known antigens of the blood. It was found that she had become sensitized in previous pregnancies to an antigen in her husband's blood that had not been encountered before. It was called the Kidd factor and tests show that about 77 per cent of the people of the United States are Kidd-positive and the remainder are Kidd-negative. Then followed the Kell factor, the Lewis, Diego, Duffy, Lutheran, Levay, and Cellano factors, all named for women who first showed antibody production against hitherto unknown antigens in their husband's red blood cells. Only in rare cases do these cause any trouble in newborn children, but they all add to the great variety of human blood antigens which are known.

TYPICAL PROBLEMS AND ANSWERS

1. A man has type A blood and his wife has type B. A child is born with type O. He accuses his wife of infidelity saying that this must mean that he is not the father of the child. Is there any justification for his suspicions on the basis of this evidence?

Answer. No, he could have been the father because both he and his wife could have carried the gene for O in the heterozygous state.

2. The man mentioned in Problem 1 still was not convinced and wanted further tests made. He was Rh-positive while both his wife and the child were Rh-negative. His blood showed no agglutination when mixed with the anti-O serum, but both his wife and child showed such agglutination. Do these tests have any bearing on the case?

Answer. The Rh tests do not have a bearing, for he could have carried the gene for Rh-negative blood. The anti-O results, however, show that he has a valid reason to question the paternity of the child. The fact that his blood did not react shows that he was homozygous $I^A I^A$ and could not father a type O child.

3. A man receives an anonymous letter threatening him with death. Police test the dried saliva on the envelope gum and find that it was sealed by a type B secretor. One suspect has this combination. He claims it is just coincidence. What are the chances that he is telling the truth?

Answer. The incidence of type B is about 10 per cent and as secretors make up about 77 per cent of the U. S. population, there would be about a 7.7 per cent chance that the suspect could have this condition by coincidence.

4. An Rh-positive man has the Fisher genotype CDe/cde while his wife has the genotype cDe/cDe. Could they have a negative child?

Answer. No! While the father could have some sperms with genes cde, all of the mother's ova would carry the gene for D. Hence, all children would be Rh-positive.

5. A baby is kidnapped and several years later a child is found who might be the kidnapped child. Blood tests of the parents and the child are given below. Could this be the lost child? If so, what are the

chances that this is the wrong child who happened to have these blood characteristics by coincidence.

	anti-A	anti-B	anti-O	anti-M	anti-N	anti-D
Mother	+	−	+	+	−	+
Father	+	−	+	+	−	+
Child	−	−	+	+	+	−

Answer. The child could not have been the child of these parents. The ABO antigen tests show a possibility and the anti-D results would have been possible, but the child is type MN and neither parent carried the gene for the N antigen.

11: GENE LINKAGE

Early in his work, Mendel discovered the principle of independent assortment. When he crossed green-wrinkled peas with yellow-round peas he obtained the recombination of green-round and yellow-wrinkled just as freely as the parental combinations. When chromosomes were discovered as the carriers of the genes, it was easy to see why this condition occurred. In meiosis there exists a random arrangement of the chromosomes during the metaphase and genes lying on different chromosome pairs will be assorted independently of their association in the parents.

Had Mendel selected two characteristics dependent upon two genes which occupied the same chromosome he would not have obtained independent assortment. Such genes tend to remain together during meiosis and are said to be linked genes. Subsequent studies of the linkage phenomena have led to valuable discoveries.

DISCOVERY OF LINKAGE

In 1903 Sutton, at Columbia University, predicted linkage by reasoning that, since the number of genes must far exceed the number of chromosomes, there would be many genes on each chromosome and these would not be expected to show free recombination.

First Demonstration of Linkage. Only three years after Sutton's prediction the first case of linkage was reported by the English geneticists Bateson and Punnett in their study of the sweet pea. In their experiments they crossed a plant that bore red flowers and spherical pollen grains with a plant that bore purple flowers and cylindrical pollen grains. A test cross of the F_1 was made and, to their surprise, yielded a ratio which was approximately 7:7:1:1 instead of the expected 1:1:1:1. The two smaller classes that were yielded represented the recombinations. This led the geneticists to suspect that the two genes involved were linked and further study showed that, indeed, this was the case. The fact that there occurred any recombinations at all, however, showed that there must be some mechanism for the separation of genes which lie on the same chromosome. We shall soon explain how this occurs.

Determining Chromosome Number by Linkage Groups. Further study of the genes of the sweet pea showed that there were seven groups of linked genes; cytologically it was found that there were seven pairs of chromosomes. In *Drosophila* four linkage groups of genes were found corresponding to the four pairs of chromosomes which can be demonstrated through studies of the cells. Thus, it became evident that it is possible to determine the number of chromosomes of an organism by genetic crosses provided that a sufficient number of inherited characteristics can be found to insure that genes from all chromosomes are represented. In actual practice both genetic crosses and cytological studies are carried out where it is deemed possible.

CROSSING OVER BETWEEN LINKED GENES

The work of Bateson and Punnett revealed that there is some method of recombination of genes that lie on the same chromosome. This process is known as crossing over and involves the actual breakage and reattachment of chromatids when they are paired during the prophase of the first division of meiosis.

Mechanism of Crossing Over. Cytological studies have yielded us a clue as to the mechanism of crossing over. During the early prophase of meiosis you will recall that the chromosomes pair and each divides so that there are actually four chromatids (tetrads) closely associated. At this time there may be simultaneous breakage and reattachment between homologous chromatids; this process creates a new association of genes. The latter condition is illustrated better than words can describe in Fig. 11.1. As the chromosomes go into the latter part of the prophase and separate from one another slightly, it is possible to see crosses between chromatids of the homologous chromosomes. These crosses are known as *chiasmata* (sing. *chiasma*) which are shown clearly in the photograph in Fig. 11.2. It is generally concluded that the chiasmata represent regions where crossing over has taken place, but there is some evidence that this may not be true for every chiasma at least. In the male *Drosophila*, for example, crossing

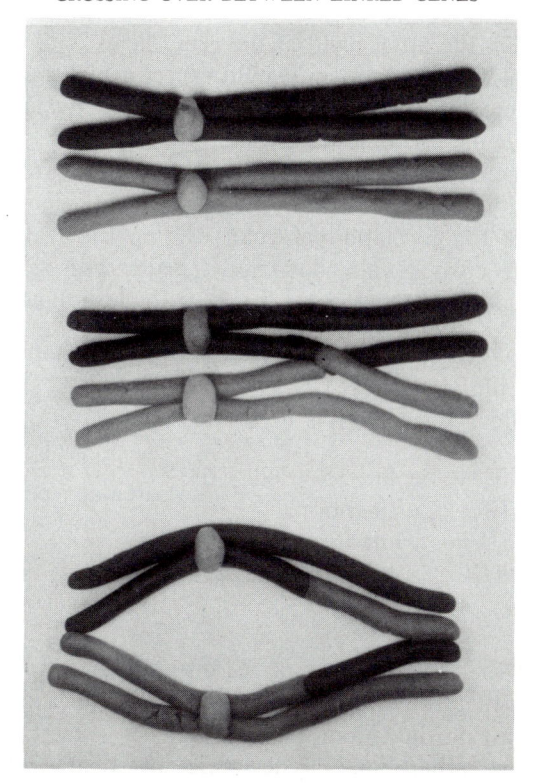

Fig. 11.1. The mechanism of crossing over as demonstrated by clay models. At the top are the two paired chromosomes in the prophase of the first division of meiosis. Each chromosome is double, but the chromatids are held together by a common centromere. In the center, it can be seen that there has been an exchange between portions of two opposing chromatids. At the bottom the chromosomes are pulling apart in the metaphase.

over does not take place under normal circumstances, yet some chiasma formation can be observed in the primary spermatocytes.

Variation in Amount of Crossing Over. The amount of crossing over may vary greatly and is correlated with the distance between the genes involved. If two genes lie close together, crossing over might occur in only a fraction of 1 per cent of the gametes produced. On the other hand, if the genes lie at opposite ends of a rather long chromosome, we would expect a high percentage of crossing over. This condition can be illustrated by two cases of crossing over in the domestic fowl.

A dominant gene (F) causes the feathers to be frizzled—they are brittle and curly and break off easily. Another dominant gene (I) inhibits the formation of color causing the feathers to be white. Hutt crossed colored-frizzled females with white-normal Leghorn males and test crossed the F_1 with the following results.

P_1 iF/iF (colored-frizzled) $\times$ If/If (white-normal)
F_1 all iF/If (white-frizzled)
test crossed with if/if (colored-normal)

Results:
iF/if	colored-frizzled	63 } parental types 80.3%
If/if	white-normal	63
IF/if	white-frizzled	18 } recombinations 19.7%
if/if	colored-normal	13
	Total	157

The fact that 19.7 per cent of the offspring are recombinations gives proof that the two genes involved must lie a considerable distance apart on the chromosome.

When two other linked genes were investigated, Taylor obtained very different results. He combined the creeper characteristic (Cr), which causes the legs to be short and the chickens to creep about, and a comb characteristic (R) which produces a rose comb while the recessive allele (r) produces the single comb. Since the gene for the creeper is a lethal characteristic when homozygous, he began with heterozygous creepers and selected the creepers from the F_1 for the test cross.

P_1 $R\,cr/R\,cr$ (rose-normal) $\times$ $r\,Cr/r\,cr$ (single-creeper)
From offspring selected: $R\,cr/r\,Cr$ (rose-creeper)
Test crossed with: $r\,cr/r\,cr$ (single-normal)

Results:
$R\,cr/r\,cr$	rose-normal	1069 } parental types 99.5%
$r\,Cr/r\,cr$	single-creeper	1104
$R\,Cr/r\,cr$	rose-creeper	6 } recombinations 0.5%
$r\,cr/r\,cr$	single-normal	4
	Total	2183

The small percentage of recombinations seems to indicate that the two genes must lie very close together on the chromosome; thus, the chance of their being separated by a crossover is very slight.

GENETIC APPLICATIONS OF CROSSOVER DATA

Through the use of the results of the crosses of linked genes we have been able to learn much about the gene arrangement on the chromosome.

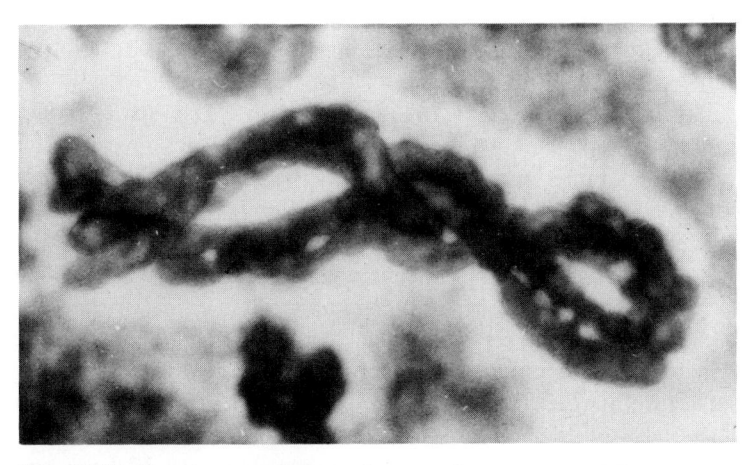

Fig. 11.2. Crossing over in grasshopper chromosomes. This is a very highly magnified photograph of a single pair of chromosomes in a living cell of a primary spermatocyte. It can be seen that each of the chromosomes is double, composed of two chromatids, and where crossing over has taken place the chromatids cross from one chromosome to the other, forming chiasmata.

Chromosome Distances as Determined by Crossing Over. Since the amount of crossing over is somewhat proportional to the distance between genes on the chromosomes, we can use the percentage of crossing over as an indication of units of distance. One per cent of crossing over is taken to indicate one unit of distance. Using the results of the cross involving feather characteristics in the domestic fowl, we can say that the gene locus which determines color or lack of color lies 19.7 units from the gene locus that determines the frizzled or normal condition of the feathers. Likewise, the gene locus for the comb characteristics lies 0.5 units from the gene locus for the creeper or normal leg condition.

Do these unit distances as determined by crossing over correspond to the actual linear distances on the chromosome? We would expect them to be the same if crossing over occurs with equal facility in all regions of the chromosome. Cytological studies reveal that there is some variation. For example, in *Drosophila*, we find that there is less crossing over in proportion to length near the centromeres and near the ends of the V-shaped chromosomes than there is in the region near the middle of each arm. Thus, we would conclude that it is easier for the chromatids to cross in some regions of the chromosomes than in others.

The Effect of Double Crossing Over. Early *Drosophila* studies on crossover percentages between two linked genes revealed that the value obtained was often less than the sum of crossover values of genes lying in between. For example, black body (b) and vestigial wings (vg) are mutant genes which show linkage with a crossover of about 17 per cent when only these two genes are used in the test. There is another mutant gene, cinnabar eye color (cn), which lies in between black and vestigial. Crossovers between black and cinnabar are about 9.0 per cent and crossovers between cinnabar and vestigial are about 9.5 per cent. The sum of the crossovers should be the distance between black and vestigial. However, this sum is 18.5 as contrasted with 17.0 per cent obtained in crosses involving only black and vestigial. Why is there disparity in the results?

The answer may be found in double crossovers. In the black-vestigial test a double crossover would restore the original order of genes and there would be no evidence of a crossover even though there had actually been two. Because of the disparity it is customary to use three genes (the ***three point cross***) for studies when the genes lie more than ten units apart. In some cases four, five, or more genes may be used. An example of the three point cross is given below.

Black-cinnabar-vestigial flies were crossed to the wild type, and the heterozygous female offspring were test crossed with black-cinnabar-vestigial males. (Only females are used from the F_1 because there is practically no crossing over in the male *Drosophila*.) The results follow:

```
+ + + (wild type)   332 ⎫ parental types 81.5%
b cn vg             326 ⎭
b + +                35 ⎫ crossovers between b and cn
+ cn vg              31 ⎭
b cn +               36 ⎫ crossovers between cn and vg
+ + vg               34 ⎭
+ cn +                4 ⎫ double crossovers 0.86%
b + vg                2 ⎭
         Total      800
```

Crossovers between *b* and *cn* (includes doubles) 72 or 9.0%
Crossovers between *cn* and *vg* (includes doubles) 76 or 9.5%
Total between *b* and *vg* (doubles counted twice) 148 or 18.5%
Total between *b* and *vg* (doubles not counted) 136 or 17.0%

From these results we can see that, had we made the cross without including the mutant cinnabar, we would have obtained a crossover distance of 17 units rather than 18.5 units because of the difficulty in detecting the double crossovers.

Interference and Coincidence. When two genes are in close proximity, we do not obtain any double crossovers between them. In other words, crossing over at one point seems to inhibit crossing over within a certain distance on either side. This inhibiting effect is known as *interference*. In *Drosophila*, interference prevents a second cross for a distance of about ten units and then gradually diminishes as the genes become further apart. The degree of interference varies in different parts of the chromosome, in different chromosomes of the same species, and in the chromosomes of different species.

We can express the degree of interference in terms of *coincidence*; thus interference is calculated by dividing the number of the obtained double crossovers by the expected frequency of doubles. We can determine the expected per cent of doubles by multiplying the percentage of crosses at the two regions involved. Using the *Drosophila* cross as an example, we obtain the following results.

Per cent crosses between b and cn	9.00%
Per cent crosses between cn and vg	9.50%
Expected doubles if there is no interference (0.09×0.095)	.86%
Expected number of doubles (.86% of 800)	7.00
Obtained number of doubles	6.00
Coincidence of interference 6/7	.86

We see that coincidence varies inversely as the degree of interference. The interference could extend from 0, which would be complete inhibition, to 1, which would yield no interference at all.

CHROMOSOME MAPPING

In organisms where a considerable number of variable inherited characteristics are known and crossover studies are possible on a large scale, it is possible to establish entire maps of the chromosomes showing the relationships of the different genes to one another. Maps have been worked out for *Drosophila*, for corn, for mice, for some of the molds and bacteria, and, to a limited extent, for man.

Mapping of Sex-linked Genes in Drosophila. We can illustrate the methods used in chromosome mapping by a case involving the X-chromosome of *Drosophila*. There exists a recessive gene for yellow body color that has been given the position of 0.0 on the chromosome, for (1) no other genes have been found on the other side of it and (2) cytological studies show that the gene is at, or very near, one end of the chromosome. Suppose then, that we have two other sex-linked genes which we would like to place properly on the map. Sex-linked genes are preferable for use in experimental crossover studies because one can allow the F_1 to breed *inter se* and study the results in the F_2 females. The genes we wish to locate are: miniature wings (*m*) and forked bristles (*f*). The results of the crossover are not arranged in any particular sequence. They are listed (see page 149) as one might have tabulated them from the actual fly counts without knowing the arrangement of the genes on the chromosome.

Before proceeding further, we must determine the proper sequence. Since the *y* is at a known location of 0.0, we know that it will be to the "left" of the other two, but which gene comes in the middle? Is the sequence *y m f* or *y f m*? The problem is most readily determined by examining the double crossovers. A double crossover will remove the middle gene from the remaining two genes and it will stand alone in one of the double crossover classes. The gene on either end will be together in the other double crossover class. The double crossovers will be expected to be the two smallest classes. When we examine the results of the crossing over, we find that number (2) and number (6) are evidently the double crossovers. Number (2) shows that miniature is in the middle class since it stands alone with the normal genes for body color and bristles. Number (6) shows that *y* and *f* are on the ends.

P_1 *y m f*/Y males × + + +/+ + + females
F_1 *y m f*/+ + + females × + + +/Y males (*inter se* cross)

Results F_2 females only:

	+ + +	26	⎫ parental types 50%
	y m f	24	⎭
(1)	+ *m f*	14	⎫
(2)	+ *m* +	2	⎪
(3)	+ + *f*	8	⎬ cross overs
(4)	*y* + +	16	⎪
(5)	*y m* +	6	⎪
(6)	*y* + *f*	4	⎭
	Total	100	

Having solved this problem, we now put the classes together in proper sequence and determine the crossover percentages.

Crossovers between *y* and *m*:

Add 1, 4, 2, and 6 = 36 or 36% of the total.

Crossovers between *m* and *f*:

Add 3, 5, 2, and 6 = 20 or 20% of the total.

Coincidence of interference:

6/7.2 = .83

Using the results of this cross we can tentatively place miniature 36 units to the "right" of yellow and forked 20 units to the "right" of miniature. The units can be placed on a chromosome map as follows:

y	m	f
0.0	36	56

If we wanted our results to be more reliable, we would have to include many more flies in our study and even consider the coincidence of interference in establishing the locations of these genes. Using such techniques, geneticists have established the extensive chromosome maps such as the one illustrated in the table on page 151, which shows only a comparatively few of the genes that have been located on the *Drosophila* chromosomes.

Pseudoalleles. In the course of study of thousands of *Drosophila*, some very interesting cases of crossing over have been found between genes which were apparently alleles. There are two genes which alter the shape and texture of the eye. They are known as Star (S) and asteroid (*ast*). For many years the genes were thought to be alleles since they seemed to occupy the same locus on the chromosome. Recently, however, it has been found that there is a crossover frequency of 0.02 per cent between these two genes, or a frequency of only one in 5,000. Since they are so close together and since they affect the eye in a similar way, it is thought that the two must have been true alleles at one time, but through some slight misalignment of chromatids in crossing over two genes of the same kind were found on one chromosome. Mutation could have caused the variation of effects. A number of such genes have been discovered in such widely divergent organisms as corn, cotton, molds, and bacteria, as well as in *Drosophila*. You will recall also from the discussion in Chapter 10 that one of the most widely accepted theories explaining the complexities of the Rh factor depends upon pseudoallelism.

Position Effect. Pseudoalleles usually express what is known as a position effect of the genes—meaning that the same genes produce different phenotypes when they are in different positions. We can illustrate this with a pair of pseudoalleles affecting eye color in *Drosophila*. As we learned in Chapter 9 the white eye series of multiple alleles is quite extensive. There is recent evidence that the genes for white and apricot might more properly be classified as pseudoalleles. When white-eyed males are crossed to apricot-eyed females, the female offspring have light apricot eyes as would be expected from the heterozygous effect of alleles. The males have apricot eyes since these genes are on the X-chromo-

LOCATION OF SOME BETTER-KNOWN GENES OF DROSOPHILA MELANOGASTER AS DETERMINED BY CHROMOSOME MAPPING

X-chromosome I	Chromosome II	Chromosome III	Chromosome IV
0.0 y—yellow body	0.0 net—net veins	0.0 ru—roughoid eyes	0.0 sv—shaven bristles
0.0 sc—scute bris.	1.3 S—Star eyes	0.2 ve—veinlet wing	0.0 ci—cubitus interruptus venation
0.6 br—broad wing	11.0 ed—echinoid eyes	19.2 jv—javelin bristles	0.0 gvl—grooveless scutellum
0.8 pn—prune eyes	12.0 ft—fat body	26.0 se—sepia eye	0.2 ey—eyeless
1.5 w—white eyes	13.0 dp—dumpy wing	26.5 h—hairy body	
3.0 fa—facet eyes	16.5 cl—clot eyes	41.4 Gl—Glued eye	
5.5 ec—echinus eyes	41.0 J—Jammed wing	43.2 th—thread arista	
6.9 bi—bifid wings	48.5 b—black body	44.0 st—scarlet eyes	
7.5 rb—ruby eyes	51.0 rd—reduced bristles	45.3 cp—clipped wing	
13.7 cv—cross-veinless wings	54.5 pr—purple eye	46.0 W—Wrinkled wing	Y-chromosome
	55.0 lt—light eye	47.0 in—inturned bristles	
18.9 cm—carmine eyes	55.9 ti—tarsi fused	48.0 p—pink eye	male fertility
20.0 ct—cut wing	57.5 cn—cinnabar eye	48.7 by—blistery wing	long bristles
21.0 sn—singed bristles	67.0 vg—vestigial wings	50.0 cu—curled wing	male fertility
27.7 lz—lozenge eyes	72.0 L—Lobe eye	58.2 Sb—Stubble bristles	(no locations worked out)
32.8 ras—raspberry eyes	75.5 c—curved wing	58.5 ss—spineless bristles	
33.0 v—vermilion eyes	100.5 px—plexus veins	59.0 Rf—Roof wing	
36.1 m—miniature wings	104.5 bw—brown eyes	62.0 sr—stripe thorax	
43.0 s—sable body	107.0 sp—speck wing	63.1 gl—glass eye	
44.4 g—garnet eyes		66.2 Dl—Delta veins	
51.5 sd—scalloped wings		69.5 H—Hairless bristles	
56.7 f—forked bristles		70.7 e—ebony body	
57.0 B—Bar eyes		90.0 Pr—Prickly bristles	
59.5 fu—fused veins		91.1 ro—rough eyes	
62.5 car—carnation eyes		93.8 Bd—Beaded wing	
66.0 bb—bobbed bristles		100.7 ca—claret eye	
		104.3 bv—brevis bristles	

Data from Bridges and Brehme.

some. In the F$_2$ the males are half apricot and half white as would be expected in a typical monohybrid sex-linked cross. E. B. Lewis, however, found that in a very large number of crosses a few red-eyed offspring appeared. He reasoned that these might be the result of crossovers and that the genes might more properly be represented as $++$ (red), $apr+$ (apricot), and $+w$ (white).

According to Lewis's theory, the two genes do not behave in the same way as other dihybrids; the position effect comes into play. The wild-type genes must be on the same chromosome in order to assert their dominant effect over the other two. This is shown below.

Genotype	Phenotype
$++/apr\ w$	wild-type red eye
$+w/apr+$	light apricot eye

Flies with these two genotypes have exactly the same gene combination, but they are in a different position with relation to one another and this alters their effect on the phenotype. (Details about the nature of gene structure and gene action are discussed in Chapter 15.)

SIGNIFICANCE OF CROSSING OVER IN SELECTION

The commercial plant and animal breeders appreciate crossing over just as much as the theoretical geneticists. When the former find some desirable characteristic linked to a characteristic which is undesirable, they need only to obtain hybrids and pick up crossovers in the second generation. Thus, the commercial breeder can select for the desirable characteristic and discard the undesirable one. The same is true of natural selection. Without crossing over there would be selection on the basis of large blocks of genes. On any one chromosome there would almost certainly be some genes which would be beneficial and some detrimental to the welfare of the organism. The sum total effect of all of the genes on the chromosome would prove to be the deciding factor in selection. With crossing over, however, it is possible for individual genes to be selected, since crossing over provides recombinations with other genes. Also, the crossing over permits selection on the basis of combination effects on different genes.

TYPICAL PROBLEMS AND ANSWERS

1. In the somatic cells of a male grasshopper there are 23 chromosomes. How many linkage groups of genes would you expect to find from genetic crosses? Explain.

Answer. There would be 12 linkage groups. The grasshopper has the XO method of sex determination—the male has only one X-chromosome, but this still represents a linkage group along with the 11 pairs of autosomes.

2. In the mouse the diploid chromosome number is 40. The gene for the waltzing characteristic is located on chromosome 10. Suppose you discover a new gene for belted body. What are the chances that it will be linked to waltzing, assuming that all chromosomes are the same length?

Answer. $\frac{1}{20}$. Since there would be a total of 20 linkage groups, the chance is one in twenty that the belted body gene will lie on the particular chromosome occupied by the gene for waltzing.

3. Suppose you cross mice having the belted body trait with mice expressing the waltzing trait. The offspring are all normal showing that both genes are recessive. You test cross the normal offspring with mice expressing the belted-waltzing trait and obtain the results given below. From these results would you think that these two genes are linked? Explain your answer.

| Belted body | 54 | Belted-waltzing | 57 |
| Normal | 49 | Waltzing | 47 |

Answer. No. These results approximate a 1:1:1:1 ratio which is expected when there is free assortment of genes in the F_1 hybrid.

4. *Drosophila* with curled wings and stripe thorax (*cu sr*) are crossed with the wild type. The F_1 (all wild type) are test crossed with curled-stripe. From the following results do you think that the genes are on the same chromosome? If your answer is yes, give the per cent of crossing over.

| Wild type | 359 | Stripe normal wings | 51 |
| Curled-stripe | 345 | Normal body-curled | 45 |

Answer. The genes are linked for the results show about a 1:1 ratio for the parental types, but the recombinations are much less in frequency. The crossover percentage is 12 (96 crossovers in 800 flies).

5. The genes *m* (miniature wings) and *f* (forked bristles) are known to be located at 36 and 56 respectively on the X-chromosome of Drosophila. You are given flies with carnation eyes (*car*), which is also a sex-linked trait. Tell where you would place it on the chromosome from the following crossover percentages.

$$\begin{array}{lll} \text{Crossovers:} & \text{between } m \text{ and } car & 26\% \\ & \text{between } f \text{ and } car & 6\% \end{array}$$

Answer. At 62.

6. Three linked genes in corn produce the following characteristics: *gl*, glossy leaves of the seedlings; *v*, virescent (seedlings are first white, then yellow and, finally, normal green); *va*, variable sterile (irregular distribution of chromosomes in meiosis). The results of a three point test cross give the following data. Determine the proper sequence of these genes and the distance between them.

+ + +	235	+ va v	4
gl va v	270	gl + v	48
gl va +	62	gl + +	7
+ va +	40	+ + v	60

Answer. Since *gl* stands alone in one of the double crossover classes, we know that it is located in the center. We recognize the double crossover classes because they are smaller (4 and 7) than any of the other classes. We find that the crossovers between *v* and *gl* represent 18.3% of the total. Between *gl* and *va* it is 13.6%. We can place these genes on a chromosome map as follows:

```
          18.3                    13.6
    |------------------|-----------------|
    v                  gl                va
```

7. What is the coincidence of interference in problem 6?

Answer.

Expected per cent (if no interference)	.183 × .136	.025
Expected number of doubles	2.5% of 726	18
Obtained number of doubles		11
Coincidence of interference	11/18	.6

8. If we assume that Fisher's theory of pseudoallelism is correct for the Rh antigens in man, do the following gene combinations indicate a position effect? Explain why you answer as you do.

Genotype	Rh factor	Rh antigens present
CDe/cde	positive	C and D
Cde/cDe	positive	C and D
cde/cde	negative	none

Answer. No. There is no indication of position effect because the C and D antigens are present when the genes *C* and *D* are on different chromosomes as well as when they are on the same chromosome. Had there been a position effect, the results would be expected to be different when the genes are not lying side by side on one chromosome.

12: CHROMOSOME ABERRATIONS

The movements and reactions of chromosomes during cell division are very exact, and in the great majority of cases there exists an orderly segregation. As in most biological processes, however, there are occasional deviations from the normal procedure and aberrant forms and arrangements of chromosomes result. As in so many scientific studies, it is such rare deviations from the normal that are of greatest value in determining the function of normal events.

ABERRATIONS INVOLVING PORTIONS OF CHROMOSOMES

In Chapter 11 we learned that in the early prophase of meiosis chromatids may become broken and reattached to their homologous mates in a mutual exchange of segments. This process is the normal and frequent occurrence known as crossing over. In comparatively rare cases the chromatids may become broken when there is no corresponding break of the homologous chromatid. A portion of a chromatid thus broken may become attached to another chromatid at a point where it has also been broken. We sometimes say that the point of breakage is "sticky," although this does not imply a literal stickiness. If there is only one break in a cell, the chromatids may become rejoined in the original positions without aberration occurring. Of course, the rejoining process may occur when there is more than one break, but additional breaks make attachment to other chromatids possible. In some cases there may be breakage at two points on the same chromatid and reattachments can occur in such a way that there is a different arrangement of genes. We shall list some of the types of aberrations which can be formed.

Deletion or Deficiency. A chromosome is said to show a deletion or deficiency if it has lost a portion of its original material.

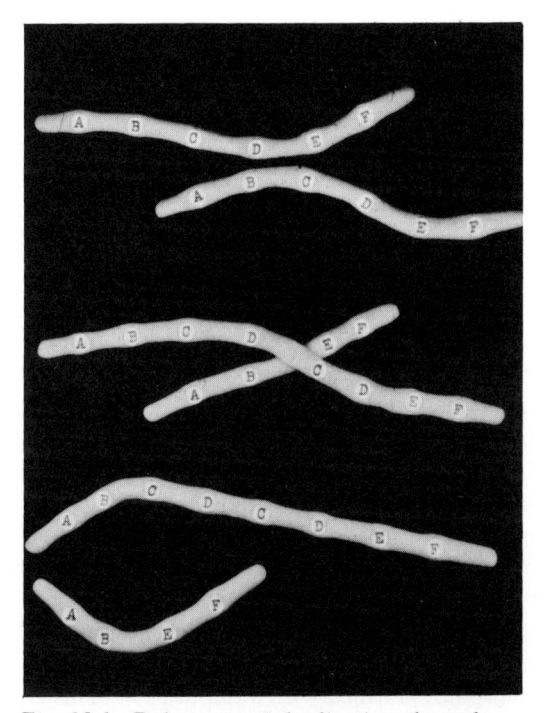

Fig. 12.1. Deletion and duplication through unequal "crossing over." Here we see clay models of two paired chromatids which break and become reattached to their homologous mates in such a way as to give a duplication of a group of genes on one chromatid and a deletion of this portion on the other.

By far the most common deletion comes about when a chromatid breaks in two places and the end portions fuse, leaving the central portion out. This is an *intercalary deletion*. A *terminal deletion* results when a portion has been lost from the end of a chromosome. Since a centromere is necessary for the normal movement of chromosomes during the anaphase, a deleted portion without a centromere will fail to get included in the nuclei which are formed and will be lost in the cytoplasm without having any genetic effects.

Deletions may be detected genetically by the expression of recessive genes in heterozygous individuals. A case in mice will illustrate this point. An autosomal recessive gene causes the waltzing trait in mice. This gene causes a defect of the inner ear which controls balance, and mice showing the trait continually go from side to side when walking—remotely simulating waltzing. In a cross of a waltzing female with a homozygous normal male, one of the offspring expressed the waltzing trait. Cytological examination of this exceptional mouse and its offspring showed that a small portion of one chromosome was missing. This obviously must have included the dominant normal allele of waltzing, so the recessive gene was expressed even though haploid.

Most large deficiencies are lethal when homozygous because the lost portion of the chromosome contains genes necessary for the normal functioning of the organism. Some cases have been found, however, in which survival is possible when the deficiencies are very small. They sometimes have an effect similar to that of a mutant gene because of the position effect on adjacent genes. In mutation studies one must be careful to check for these small deletions when an apparent mutation has been found.

Duplication. Frequently a portion which becomes deleted from one chromatid becomes attached to the homologous chromatid and a chromosome is formed with a duplication. This chromosome has two sets of genes for the duplicated portion. Such a combination deletion-duplication could occur if crossing over took place when the chromatids were not exactly aligned during their synapsis. We have already learned how such duplication of small portions of the chromosome could account for the origin of pseudoalleles. Through mutation some of the duplicated genes could become changed from their original form, thus adding to the gene pool of the species.

Inversion. Sometimes a chromatid may break in two places and then become reattached in the same region, but in reverse order. This situation might occur when the chromatid forms a loop in the center. There is no change in the genes on the chromosome formed, but some of the genes will be in inverse order. The occurrence of this inversion has one very important genetic significance. Normal synapsis will not occur when such a chromosome pairs with a normal chromosome in the early prophase of meiosis. There seems to be a repulsion at that region where the genes do not match (this condition can be seen cytologically). Since there is no synapsis at the inverted region, there will be no

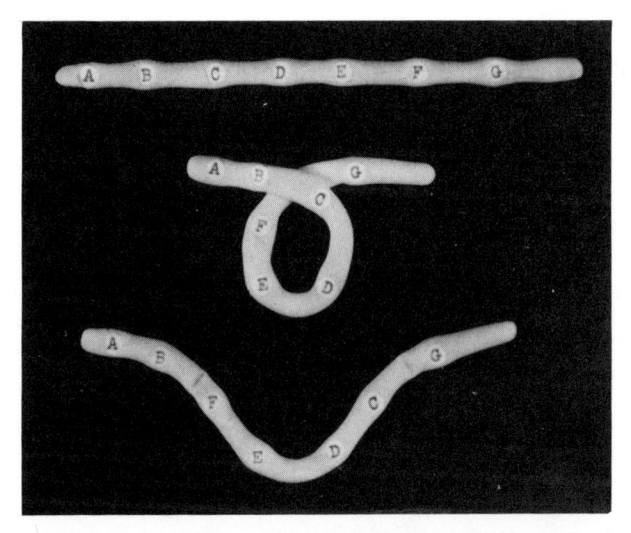

Fig. 12.2. *Inversion. A breakage and reattachment of the center portion of a chromosome can result in a region in which the genes are inverted with respect to their original order.*

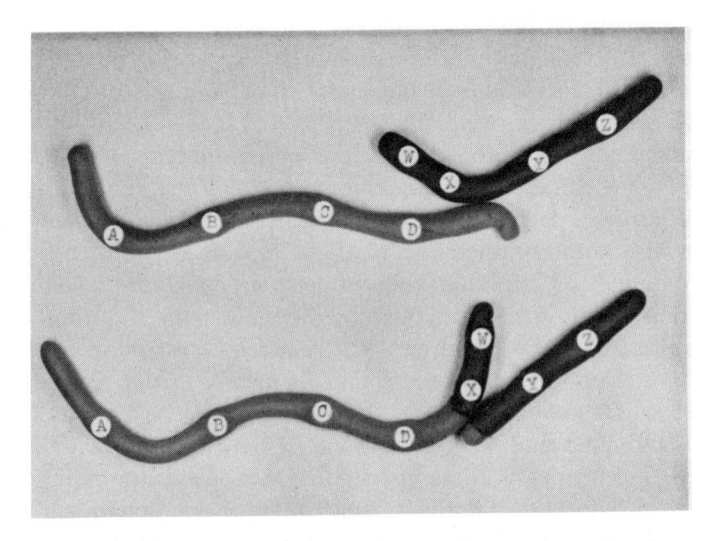

Fig. 12.3. *Translocation. An exchange of a portion of a chromosome with a non-homologous mate causes the transfer of genes from one linkage group to another.*

crossing over. When geneticists wish to maintain genes on a chromosome in the heterozygous state and not have recombinations through crossing over, they may purposely employ chromosomes containing inversions.

Translocation. This type of aberration involves the exchange of portions by different chromosomes. Thus, a block of genes from one chromosome may shift to a new linkage group on another chromosome. It appears as if this process is always accompanied by a simultaneous shift of some of the genes in the reverse direction. The entire procedure is known as a *reciprocal translocation* and is illustrated in Fig. 12.3. The portion shifted may be very unequal in size, but it appears that there cannot be an attachment of one portion unless there is at least a small portion exchanged in the process. An unbroken end of a chromatid does not seem to have the necessary affinity for attachment.

ABERRATIONS INVOLVING ENTIRE CHROMOSOMES

Sometimes abnormal events during cell division may cause the loss or addition of entire chromosomes, and this condition may have pronounced phenotypic effects. Duplications or subtractions of one or several entire chromosomes are known as aneuploids, while similar aberrations involving entire haploid sets of chromosomes may result in polyploids.

Aneuploids. The process known as non-disjunction may result in (1) some gametes with two chromosomes of one kind and (2) other gametes lacking this chromosome. When such gametes unite with normal gametes we have one type of zygote with three chromosomes of one kind and another with only one. **Non-disjunction** occurs when the two chromosomes adhere during the metaphase of meiosis and are not separated—both move to one pole. This condition has been found frequently in the X-chromosomes of *Drosophila* females. It results in some eggs with two X-chromosomes and some with none. When fertilized with normal X and Y sperms these aberrant gametes yield interesting results which in turn aid the genetic study of sex determination. (See Chapter 13 for further discussion.) Cases in *Drosophila* have been discovered in which such non-disjunction occurred in every meiosis. Cytological examination shows that this situation is due to the attachment of the two X-chromosomes at one end and the inability to separate during meiosis. The aforementioned are known as *attached X-chromosomes* and have genetic value because sex-linked characteristics present in a male will all be transmitted to his male offspring (as shown in Fig. 12.4).

Fig. 12.4. Attached X-*chromosomes in* Drosophila. *At the lower left it can be seen that there are two X-chromosomes attached at the centromere end. These do not separate in meiosis and this results in 2 X eggs, which produce females when fertilized with Y sperm. The Y-chromosome is shown in the lower right.*

A. F. Blakeslee made an extensive study of the Jimson weed, *Datura*, and collected data on numerous cases of aneuploidy. The normal form of this plant has 12 pairs of chromosomes, but variations of many kinds exist. Some plants had three of one kind of chromosome and normal 11 pairs of the others—this type of plant is known as *trisomic*. Some plants were haploid with respect to one chromosome and diploid for the others—this type is known as *monosomic*. Some plants were found to have two or more extra chromosomes—this type is known as a *polysomic*. The variations in *Datura* were associated with specific phenotypic effects; it is possible to recognize the particular chromosome aberration by the appearance of the plant.

ABERRATIONS INVOLVING ENTIRE SETS OF CHROMOSOMES

Whenever aberrations occur which involve entire haploid sets of chromosomes, we speak of the condition as polyploidy.

Formation of Polyploid Cells. Abnormal mitosis can give rise to cells which have double the normal chromosome number. Mitosis may start and the chromosomes become duplicated, but then the process stops and the cell is left with the *tetraploid* rather than the diploid number of chromosomes. Later, normal mitosis may take place and many cells may be produced with the double chromosome number. This condition is of special significance in plants because a tetraploid cell formed in the growing tip of a stem can give rise to an entire branch which has this type of cell. This situation can have great value because a tetraploid variety may produce a larger and more valuable commercial product—tetraploid apples may be about twice the size of the diploid variety.

Polyploidy can also result from abnormal meiosis in which the chromosomes pair and duplicate, but in this case there exists nondisjunction of the entire group and they all move to one pole. This later condition results in some diploid gametes which, when united with normal haploid gametes in fertilization, give offspring with the *triploid* chromosome number. The triploid number is an unstable one and triploid organisms are usually sterile, but in plants the organism can be maintained by the rooting of cuttings and by grafting.

Induction of Polyploidy. A number of chemicals have been found which inhibit normal mitosis. When applied to cells in the proper concentration the chromosomes divide and reach the metaphase state, but there is no normal spindle figure and they go no further. This process can be seen under a microscope. Tissue chemically so treated will show many more metaphases than normal, but there will be no anaphases or telephases. In time the cells return to the interphase, but the chromosome number is tetraploid. Later, the treated cells may go through normal mitosis and produce more tetraploid tissue.

The common chemical used for this purpose is known as *colchicine,* a highly poisonous compound which is used in minute quantities as a treatment for gout in man. This chemical can even be applied to tetraploid tissue and yield *octaploid* cells with eight haploid sets of chromosomes. Some cells have been reported with even higher multiples of the haploid number. This chemical technique has commercial possibilities because improved varieties of plants can sometimes be produced in this way. In higher animals it is of less value because animals do not have terminal growth points nor asexual means of propagation. Experiments have shown, however, that the primary oöcytes of rabbits treated with colchicine have resulted in diploid eggs. When the latter are fertilized with normal sperms and transplanted into the uterus of females, triploid offspring are produced.

Cytologists often use colchicine in chromosome experimentation in order to obtain a much larger number of cells showing the chromosomes during mitosis. Since mitosis goes to the metaphase and stops, it is possible to build up mitotic figures over a period of several hours. This operation is especially valuable in tissue where mitosis is not very frequent. In human chromosome studies using tissue cultures, the colchicine technique has been used extensively, and its application to the solution of one of the problems of sex determination in man is discussed later (see Chapter 13).

THE SALIVARY CHROMOSOMES OF DROSOPHILA

Cytological investigation of chromosomal aberrations has been difficult in many organisms because of the size of the chromosomes. They are frequently too small to show details needed for best understanding. Since it has been discovered that the chromosomes of the salivary glands of *Drosophila* are truly giant in size, geneticists have an excellent means of studying chromosome aberrations in detail. It is fortunate that such chromosomes are found in a form which is as extensively investigated as *Drosophila* because this situation permits the correlation of cytological and genetic findings.

Appearance of Salivary Gland Chromosomes. The larvae of *Drosophila* are voracious feeders and are well equipped for such

164 CHROMOSOME ABERRATIONS

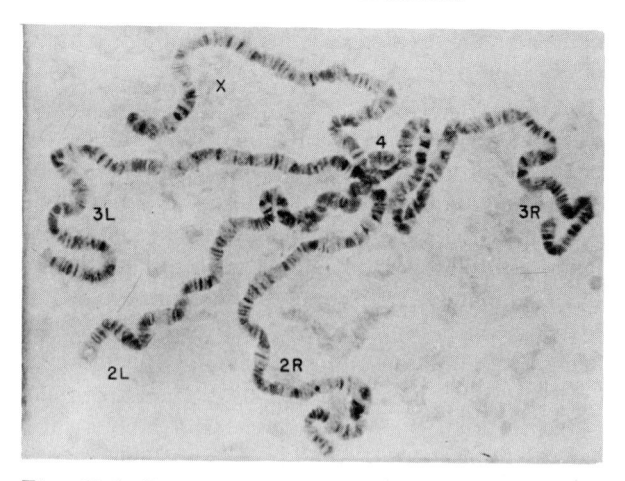

Fig. 12.5. The salivary gland chromosomes of Drosophila melanogaster. This is a photograph of a smear of the glands showing the five long arms coming out from the chromocenter. The origin of the arms is identified by the numbers and letters. (2 L is the left arm of the second chromosome, etc.) (From B. P. Kaufmann, Carnegie Institute.)

feedings by the presence of a large pair of salivary glands. The cells of the salivary glands are extremely large and the chromosomes within them are likewise of a giant size in comparison to other type of chromosomes. The glands can easily be removed and smeared on a microscope slide for study. When viewed under the microscope, chromosomes can be seen which are a hundred times as long and much thicker than those found in the other somatic cells. In addition, the chromosomes show an unusual pattern of finely detailed bands. The latter are characteristic for the different chromosomes and a small portion of the chromosome can be identified by the pattern of bands which it bears.

When we count the chromosomes in one cell, we notice yet another strange detail. There appear to be only five chromosomes and these are all attached at one end to a central mass (the *chromocenter*). *Drosophila melanogaster* has eight chromosomes in one diploid cell; whence come the five? Further study reveals the answer. Sometimes a slight separation running lengthwise can be seen on a chromosome which has been subjected to considerable pressure on the slide. An analysis of this condition has found that (1) the chromosomes are actually paired in **somatic synapsis**, so each arm seen is actually two closely paired chromosomes and that (2) the chromocenter represents the centromeres of the chromosomes together with a portion of each chromosome that does not contain genes (the *heterochromatin*). The gene-containing portion of a chromosome is known as *euchromatin.*

With this knowledge we can analyze each of the five arms that radiate out from the chromocenter. The X-chromosome with its terminal centromere yields one arm—this will be only one half as thick as the other arm if the larva is male because it will not be composed of two synapsed chromosomes. The Y-chromosome, composed mostly of heterochromatin, is to a great extent located

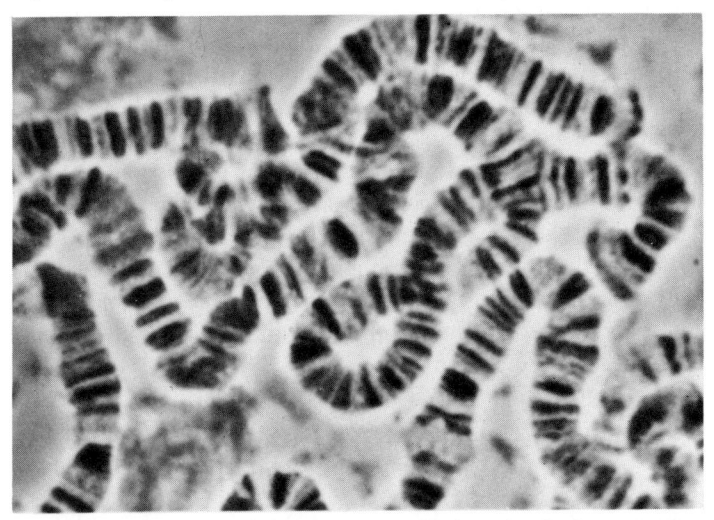

Fig. 12.6. Close up of Drosophila *salivary gland chromosomes showing details of the banding. This is a photograph of a fresh unstained smear made with a phase microscope.*

in the chromocenter and therefore does not manifest itself clearly. The second and third chromosomes have central centromeres; hence each has two long arms extending out from the chromocenter. This situation accounts for the five arms, which are easily seen. The fourth chromosomes are very short and are usually overlooked, but sometimes they can be seen as a little bump on the chromocenter.

Origin of the Giant Chromosomes. In the great majority of cells the interphase chromosomes are in the form of very long and very slender thread-like bodies with small minor coils. During the prophase of mitosis the chromosomes duplicate, thereafter becoming much shorter and thicker through the process of continued coiling and matrix deposition around the coils (see Chapter 3). In the salivary glands of the larvae of insects in the order *Diptera*, however, the chromosomes duplicate but do not shorten by further coiling; there is no cell division—the cells get larger, but do not divide. This process is followed by further duplications until there are chromonemata by the hundreds in these paired chromosomes. This condition furnishes the chromosomes with their thickness; the fact that they are not tightly coiled gives them their great length.

The bands on the chromosomes which appear more dense under the phase microscope (dark phase) and which stain more heavily with the commonly used nuclear dyes represent areas of heavy deposition of nucleic acid (DNA). Since we know that this is the material of which genes are made, it might be postulated that the bands are the areas of gene location while the light areas in between the bands are heterochromatic areas. This possibility has led to the theory that each band represents the area of one gene, and calculations of the number of genes on one chromosome as compared to the number of bands which can be counted are in rather close agreement.

VALUE OF GIANT CHROMOSOMES IN ABERRATION STUDIES

Once a cytogeneticist has learned the pattern of bands on the chromosomes of a species of *Drosophila*, he can easily detect any chromosome aberrations because the bands will be in an abnormal arrangement. This situation has great value in identifying the nature and extent of aberrations.

Locating Genes through Small Deletions. It is possible to determine the particular band which is associated with a particular gene through small overlapping deletions. Since there exists somatic pairing, a normal chromosome will express a small buckle when it is in the same cell with a chromosome bearing a small deletion. There may be only a few bands on the buckled portion, yet genetic studies show that a particular recessive gene lies in this region because the gene will be expressed in the heterozygote. Studies of several such deletions can locate the gene on a particular band.

Inversions and Translocations. Inversions can be detected easily on smears of a salivary gland chromosome. When a chromosome with an inverted portion is paired with a normal chromosome, there will be a region of repulsion where the genes do not match; this process will cause the parts to be separated for the distance of the inversion. A photograph of such an inversion is shown in Fig. 12.7. Translocations show up with equal clarity. There will be a pairing of the homologous portions even though they are on different chromosomes. This condition yields some interesting cross-connected patterns between chromosomes.

Duplications. The bar-eye locus on the X-chromosome reveals some interesting facts about small duplications. In heterozygous

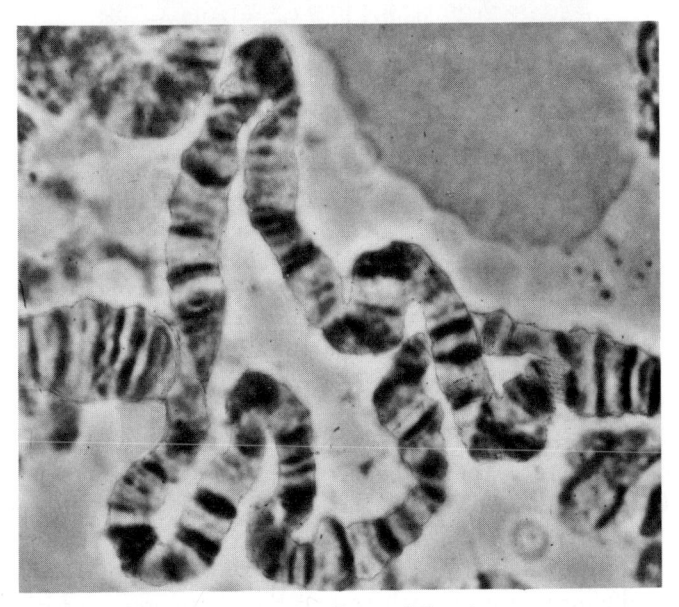

Fig. 12.7. An inversion of a Drosophila chromosome causes a repulsion when paired with a normal chromosome. The inverted region can be clearly seen because the chromosomes are not paired while the pairing is very close at either end of the inversion. (Phase photograph of fresh, unstained smear.)

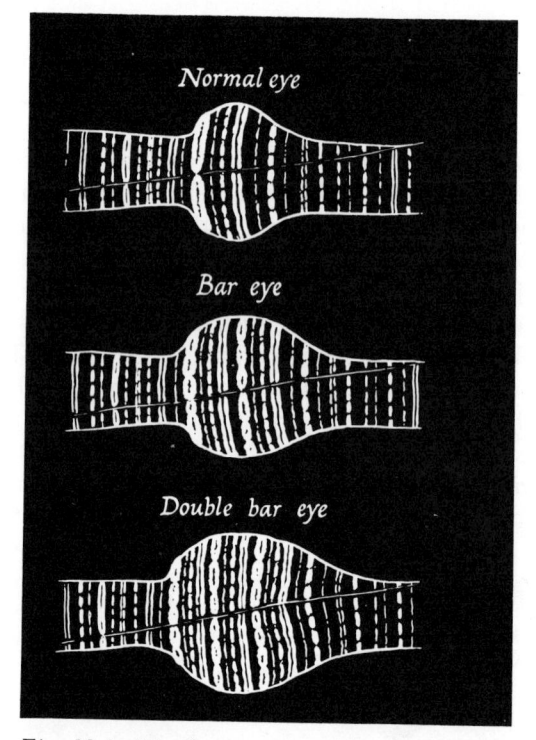

Fig. 12.8. The bar-eye locus of the X-chromosome as seen in salivary gland smears. It can be seen that bar eye is actually due to a duplication of a portion of the chromosome probably due to unequal crossing over. Another unequal crossing over can result in double bar eye. (From Winchester, Genetics, *Houghton Mifflin.)*

females, the eye is bar-shaped. In homozygous females or hemizygous males, the eye is even narrower. At first sight, one would think that this condition was a characteristic due to an intermediate gene, but salivary gland smears show that it is actually due to duplication of a small part of the chromosome which includes four bands. The condition probably arose through unequal crossing over in this region. Sometimes, further unequal crossing over in homozygous bar-eyed females will occur and one chromosome will result with three sets of these four bands. This process causes the ultra bar eye—an extremely narrow slit of eye tissue. The other chromosome will have lost the extra four bands and now, with only one set of the four bands, it produces a normal eye. Thus, we can conclude that bar eye is actually due to a position effect of this portion of the chromosome. Two sets of these four bands on one chromosome in a male cause bar eye, but in a female with two normal X-chromosomes there will also exist two sets of these bands; thus the eye is normal because there is only one set of bands on each chromosome. There must be two sets on one chromosome to cause the bar eye.

The salivary gland chromosomes therefore yield a comparatively easy solution to many genetic problems. It is unfortunate that more species do not have such giant chromosomes.

13: THE ROLE OF CHROMOSOMES IN SEX DETERMINATION

Chromosome aberrations have enabled us to obtain a much clearer picture of the role of chromosomes in the determination of sex (see Chapter 7). In this chapter we shall learn how abnormal distributions of chromosomes and parts of chromosomes have given us valuable information about the normal function of chromosomes in sex determination.

DROSOPHILA

Details of the mechanism of sex determination were first worked out on *Drosophila melanogaster*; thus, geneticists have an almost complete picture of the role of the chromosomes in this form.

Sex Chromosomes and Autosomes. Since flies with two X-chromosomes are female and those with one X and one Y are male, it would appear logical that the Y-chromosome is an important element in determining that the genes for male sex shall be expressed. Experimental evidence, however, shows that this is not the case in *Drosophila*. In rare cases, during the first meiotic division of spermatogenesis there is non-disjunction of the sex-chromosomes and they both move to one cell. This condition causes some sperms to have both an X- and a Y-chromosome, while other sperms have neither. When an XY sperm unites with a normal egg, a zygote is formed with the normal three pairs of autosomes, two X-chromosomes, and a Y-chromosome. Such a fly is female, perfectly normal in appearance and function. This seems to rule out the Y-chromosome as a male-determining agent.

Further support for this concept is found when a sperm carrying no sex chromosomes fertilizes a normal egg. This situation yields a zygote with three pairs of autosomes and one X-chromosome. The process forms a male, perfectly normal in appearance, but in reality sterile. Thus, it appears that the Y-chromosome must have genes which are functional in normal male fertility, but not in the production of the other detectible male characteristics.

The Ratio Theory of Sex Determination. The aforementioned evidence indicates that it is the presence of two X-chromosomes which determines that a fly shall be female whereas a single X causes it to be male. C. B. Bridges of Columbia University worked

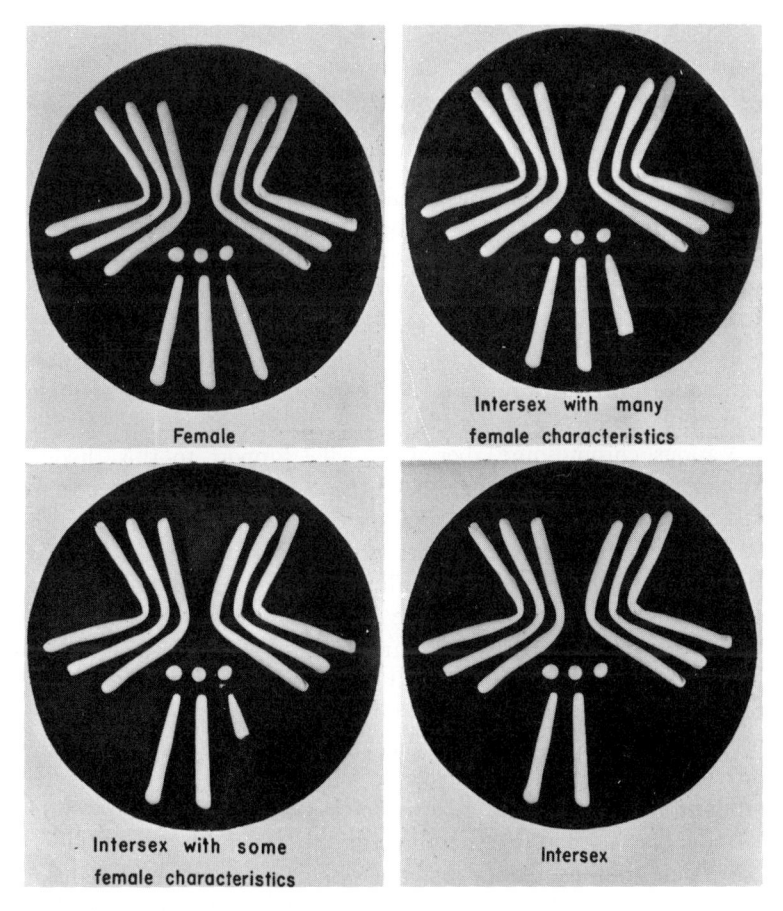

Fig. 13.1. Evidence that female-determining genes are on various parts of the X-chromosome. The triploid fly with three X-chromosomes is female in accordance with the ratio theory. When portion of one of the X-chromosomes are missing there is a tendency toward intersexuality depending upon the size of the missing portion. (From work of Dobzhansky and Schultz.)

out an interesting and plausible theory to explain how this situation could arise. Keep in mind the fact that there are genes for both sexes in all of the cells and sex determination is brought about by some trigger that allows the genes of one sex to be expressed while those of the other sex are suppressed. Bridges' experiments indicate that it is the ratio of the X-chromosomes to the autosomes which triggers the reaction.

The X-chromosome appears to carry genes which trigger the female expression, whereas the autosomes appear to carry genes which trigger the male expression. The number of autosomes remains constant, two of each kind, but the X-chromosomes vary. We might set an arbitrary value of 1.5 in female determination for each X-chromosome and a value of 1.0 in male determination for an entire haploid set of autosomes (three) and represent this as A. Since we have shown that the Y-chromosome does not play a part in this determination, it would have a value of 0.

A normal female would be expressed as XXAA and this condition would yield a 3:2 ratio in favor of female expression. A normal male, however, would be 1.5:2 with the ratio in favor of male expression.

Various chromosome aberrations give support to this theory. When there is non-disjunction of the X-chromosomes in oögenesis, eggs will be formed with either two X-chromosomes or with none. The various results of fertilization by the X and Y sperms are shown in the table below. Flies which have triploid chromosome numbers express even more convincing evidence in favor of the theory. When a fly has three sets of autosomes and two X-chromosomes we can see that the ratio would be 3:3. This equality of male-determining and female-determining elements results in an intersex—the sex characteristics are approximately intermediate. The following table shows that it is possible to produce a supermale and a superfemale when the ratios are more extreme than normal in either direction.

Sex-determining Genes on the X-chromosome. Dobzhansky and Schultz, of the California Institute of Technology, extended the information on this theory by studying triploids with various fragments missing from one of the X-chromosomes. A fly with three of each kind of chromosome would be female, but when a portion of one of the X-chromosomes was missing there was a tendency toward the intersexual condition—the greater the size

Egg	Sperm	Zygote	Ratio male : female	Sex
AX	AX AY AXY	AAXX AAXY AAXXY	2:3 2:1.5 2:3	female male female
AXX	AX AY	AAXXX AAXXY	2:4.5 2:3	superfemale female
AAX	AX AY	AAAXX AAAXY	3:3 3:1.5	intersex supermale

A Haploid set of autosomes, male value of 1.
X X-chromosome, female value of 1.5.
Y Y-chromosome, value of 0.

of the missing portion, the greater the intersexual tendencies. This condition is illustrated in Fig. 13.1.

HYMENOPTERA (*Habrobracon*)

The ratio of X-chromosomes to autosomes obviously cannot be the explanation for many of the insects which have the same ratio of chromosomes in both sexes. The honeybee males, for example, are haploid and the females are diploid, but the ratio of chromosomes to one another is the same in both. A theory to explain the condition in the *Hymenoptera* (insects with membranous wings) has been worked out by Whiting at the University of Pennsylvania. Using the small parasitic wasp, *Habrobracon junglandis*, which has sex determination like that in the honeybee, Whiting discovered that diploid organisms are normally female and haploid organisms are normally male. He noted that there appears to be a series of sex genes which have a large number of multiple alleles. These are designated as x^a, x^b, x^c, etc. When any two alleles are heterzygous in a zygote, a female is produced. The unfertilized eggs always develop into males since it is impossible to have two genes at the same locus in a haploid zygote. Diploid wasps, on the other hand, are found to be almost exclusively females—there are so many multiple alleles that they would almost always be heterozygous for the sex genes.

Whiting discovered, however, that it was possible to obtain

some diploid males. He accomplished this through close inbreeding, thus achieving a high degre of homozygosity in the cultures. As an example of how this occurs let us assume that a female is heterozygous x^p/x^r, with the x^p from the maternal parent and the x^r from the paternal parent. If she were crossed back to the male parent she would produce some eggs with the homozygous genotype x^r/x^r—these would be diploid males. In addition, some of her male siblings (brothers) would carry the x^p and she could, through mating with these, produce some homozygous diploid male offspring. These diploid males are highly sterile, however, and would almost never function in reproduction in the rare cases where they might be produced in a natural environment.

CHROMOSOMES AND SEX IN ZW ANIMALS

In animals with the ZW method of sex determination there is some variation from the sex-chromosome-autosome balance as found, for example, in *Drosophila*.

The Gypsy Moth, Lymantria. Richard Goldschmidt, a German-trained geneticist associated with the University of California, studied the ZW form in great detail. He found that it was possible to produce intersexes by crossing moths collected from different parts of the world (for example, crossing a Japanese male and a European female). There is a distinct sexual dimorphism in these moths and it was thus possible to detect degrees of intersexuality easily. Some intersexual individuals had the female (ZW), constitution and began developing as females, but later developed male traits, sometimes even completely converting to the male morphology. Others had the male (ZZ) constitution and began developing as males with female traits appearing later.

Goldschmidt proposed the theory that this converting tendency could be explained if we assumed that female-determining factors were located in the cytoplasm and male-determiners were on the Z-chromosome. The intersexes arise because geographically separated races have varying degrees of strength in the sex-determining factors. Thus, a ZW moth with a "strong" Z-chromosome from the paternal side and a "weak" cytoplasm from the maternal side would start life as a female and then develop the male tendencies as the cytoplasm had its effect. A more recent theory proposes that the female determiners are on the W-chromosome rather than in the cytoplasm. This factor could account for the "strong" and "weak" races and is more in line with genetic studies dealing with other species.

Birds. The somewhat limited work on birds indicates that the female-determining genes are located on the W-chromosome and the male-determining genes are on the Z-chromosome. Thus, it is the relationship between these two which determines the sex. A single W would then be assumed to have sufficient strength to overcome the male-determiners on a single Z.

PLANTS WITH SEPARATE SEXES

Most plants are monoecious and, therefore, encounter no problem in their sex determination process, but throughout the plant kingdom there are some plants which are dioecious and for these there must be some other method of sex determination.

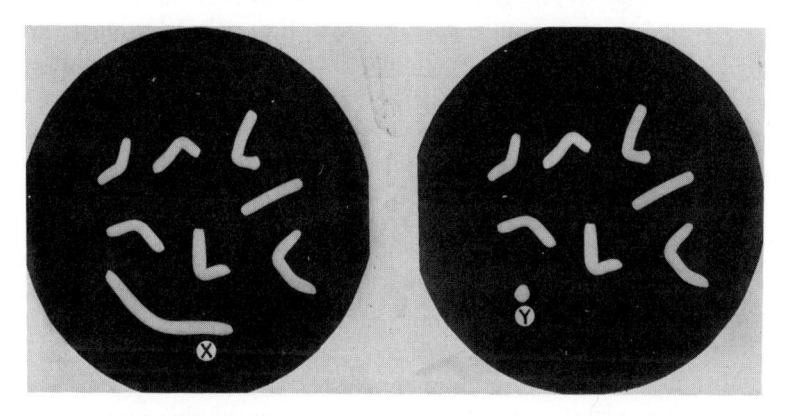

Fig. 13.2. Sex-chromosomes in the liverwort. The haploid cells of the female liverwort show a rather large X-chromosome while those of the male show the small Y-chromosome.

Haploid plants. In some of the simple plants, in which the dominant generation is haploid, the problem of sex determination is rather easy to solve. The liverworts have been worked out very nicely and are a good example. In typical liverworts the sperms are borne on stalked antheridia on male plants while the eggs are borne on stalked archegonia resembling miniature palm trees on female plants. The sperm swims to the egg and a diploid zygote is formed, but after only a few mitotic divisions there is a reduction division and haploid spores are formed. Each of these may grow into a liverwort, producing both male and some female plants.

A chromosome study of the liverwort, *Sphaerocarpus*, by Allen, showed that there were fourteen autosomes, an X-chromosome, and a Y-chromosome in the diploid tissue. After reduction, the spores contain seven autosomes and either an X- or a Y-chromosome. The X is easily recognized since it is much larger than the Y. Spores with the X develop into female plants and those with the Y become males. It is easy to see that this must be a case where the female-determiners are on the X and the male-determiners are on the Y.

Diploid Plants. Of the diocious, diploid plants that have been studied, most appear to have the XY method of sex determination, although a few have been found with XO and ZW methods.

A study by Westergaard and Warmke on *Melandrium*, a seed plant in the pink family, revealed that the male-determiners apparently occupy the Y-chromosome. There are 24 chromosomes in the species studied and these are located in 12 pairs in the female plants—the males have 11 pairs and the unequal X and Y. Occasionally, tetraploid plants are found which are one of three types: those with 44 autosomes and 4 X-chromosomes, which are females; those with 44 autosomes, 2 X-chromosomes, and 2 Y-chromosomes, which are males; those with 44 autosomes, 3 X-chromosomes, and one Y-chromosome, which are also males. These results indicate that the Y-chromosome carries the male-determiners and its presence will evoke the male characteristics even though it is outnumbered by 3 X-chromosomes.

Triploid plants reveal further confirmation of this theory. The triploids may be produced by crossing tetraploid plants with the normal diploid. This operation produces 33 plus XXX females, 33 plus XXY males and 33 plus XYY males. The most convincing evidence comes from these polyploid plants that contain fragments of the Y-chromosome. Intersexes are produced when approximately one half of the Y-chromosome is missing.

MAN

Chromosome studies in man have always been complicated by the difficulty of obtaining suitable tissue which would show the chromosomes clearly. As a result, it has frequently been necessary to postulate human relationships from the results obtained in studies of experimental animals. Recent developments in tissue culture techniques, however, have opened up new fields of investigation, and geneticists by such methods have learned much about human chromosomes, including their relationship to sex. It is possible to remove a small quantity of blood and treat it with an extract of the kidney bean. This extract destroys the red cells and stimulates the white cells to divide. They can be cultured for a time on a special tissue culture medium so that a large proportion of the cells will continue to divide. A smear of these cells is then made on a microscope slide and the chromosomes can be seen clearly under the microscope.

Klinefelter's Syndrome. This syndrome is an abnormal sex condition in which an apparent male has underdeveloped sex organs and is sterile; there is noted some degree of breast enlargement, the fat deposits and hair growth are female in nature, and there are various glandular defects.

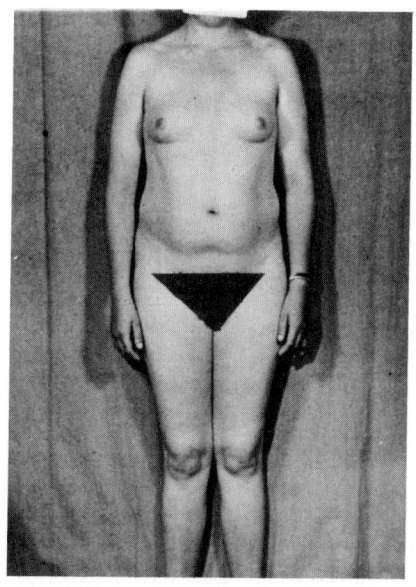

Fig. 13.3. *Klinefelter's Syndrome. This person is a male, but the reproductive organs are underdeveloped and, as can be seen, there is some degree of development of female characteristics. Chromosome studies of this and similar individuals reveal that the sex chromosome constituents of the cells are* XXY. (*Courtesy of C. Povl Riis, Copenhagen County Hospital.*)

In the past it has been postulated that such persons might be genetic females with some hormone unbalance which causes a considerable degree of sex reversal. Using the previously described technique of dividing cell cultures, however, it has been found that these individuals possess the XXY sex-chromosome combination. Such a combination could be brought about by non-disjunction in the gametogenesis of one of the parents.

Turner's Syndrome. A person expressing Turner's syndrome is a phenotypic female, but the female characteristics are underdeveloped and the reproductive organs never reach functional maturity. Chromosome studies of tissue culture cells show that such individuals have the XO sex-chromosome constitution. An interesting genetic confirmation of this has been discovered. There are cases on record of color-blind individuals with the Turner's syndrome. In all of these cases the father was found to have normal vision. Since color blindness is a sex-linked trait which is recessive, a female should never express it unless the father expresses it. The results in this case show that the X-chromosome bearing the gene came from the mother—the sperm contained no sex chromosomes, probably because of non-disjunction in spermatogenesis.

Location of Sex Determiners. The results of the aforementioned studies offer convincing evidence that the X-chromosome in man bears female-determiners and the Y-chromosome bears male-determiners.

Autosomal Non-disjunction. Non-disjunction of autosomes as well as of sex-chromosomes is known to take place in man. In cases of the consequent triploid or haploid condition of the larger autosomes, the effect is lethal, but triploids of some smaller autosomes can survive. The most common, and best known, of these is the triple-21. A person receiving three of the small, number 21 chromosomes has a condition known as Down's syndrome, or Mongolism as it is more commonly known. About one child out of each 600 born has this abnormality. The extra chromosome 21 causes retarded physical, sexual, and mental development. The afflicted individuals are on the idiot or imbecile level of intelligence.

14: GENE STRUCTURE

The gene has been the object of detailed study. Through experimental breeding and cytological investigation, geneticists have learned much about gene locations on the chromosomes and the ultimate effect of genes on the organisms which carry them. During recent years man's eternal curiosity about the unknown has led to more intensive research into the physical and chemical nature of genes. Such research has produced rewarding data, and in this chapter we shall survey some of the findings.

THE CHEMICAL NATURE OF GENETIC MATERIAL

When chromosomes are analyzed by microchemical techniques it can be shown that they are composed of nucleoproteins—proteins closely associated with *deoxyribonucleic acid (DNA)*. The DNA is found only in chromosomes—it must be synthesized there from other components in the cell—and, therefore, must be an important part of the genes. It was once postulated that the genes are composed of nucleoproteins; genetic diversity was thus plausible because of the many variations possible for the protein component. Geneticists did not see how there could be present sufficient diversity of the DNA to yield the numerous varieties of genes which exist. Modern research, however, indicates that the DNA is apparently the true genetic material and that an infinite diversity is possible. We have evidence to support this viewpoint from a number of sources.

Evidence from Bacterial Transformation. For many years bacteria were neglected as subjects for genetic investigation because of the paucity of physical variations which could be observed. It has been found, however, that bacteria exhibit a great diversity in their measurable physiological characteristics. In addition, most bacteria appear to be haploid and hence express all of the genes which they carry. This factor, coupled with their very short life cycle and the fact that physiological variants can be easily isolated, has brought bacteria to the forefront of organisms now employed in genetic research. Part of that research has led to the discovery of the principle of transformation which offers strong evidence that DNA is the genetic material.

We can illustrate the principle of transformation by an experiment with the bacterium, *Pneumococcus*, which is associated with bacterial pneumonia. The typical growth of these bacteria on laboratory culture media is in the form of smooth-surfaced colonies. A mutant gene causes the colonies to grow with a rough surface. Microscopic investigation reveals why this is the case. The smooth colonies are formed of bacteria inclosed in a gelatinous type of capsule which blend to produce the smooth surface. Bacteria taken from a rough colony, however, are lacking in capsules. This is important medically because the rough colonies are not sufficiently virulent to cause pneumonia.

In the transformation experiment some bacteria from a smooth-surfaced colony were broken up and an extract from them was placed in a food medium upon which unencapsulated bacteria were grown. A few of the colonies which resulted from this procedure were of the smooth type, demonstrating that in some way the gene for capsules had been transferred to a number of the bacteria through the food upon which they grew. Mutation was ruled out, for the transformations were ten thousand times more abundant than if a mutation had caused them. When the extracts were so refined that they were almost pure DNA, the number of transformations remained just as great. When the entire nucleoproteins were used there was no increase in the transformations, and thus it has been concluded that DNA is the genetic material.

Evidence from Virus Studies. The viruses, the smallest units of matter which may be classified as living—provide an opportunity to study the chemical nature of genes at the most elemental level. Chemical analysis and electronphotomicrography reveal that the virus particle is composed of nucleoprotein, the same as the chromosomes of higher forms of life. There appears to be an outer protein coat encompassing an inner core of nucleic acid. In most cases the nucleic acid is DNA, although there are a few viruses, such as the tobacco mosaic virus, which contain the closely related RNA.

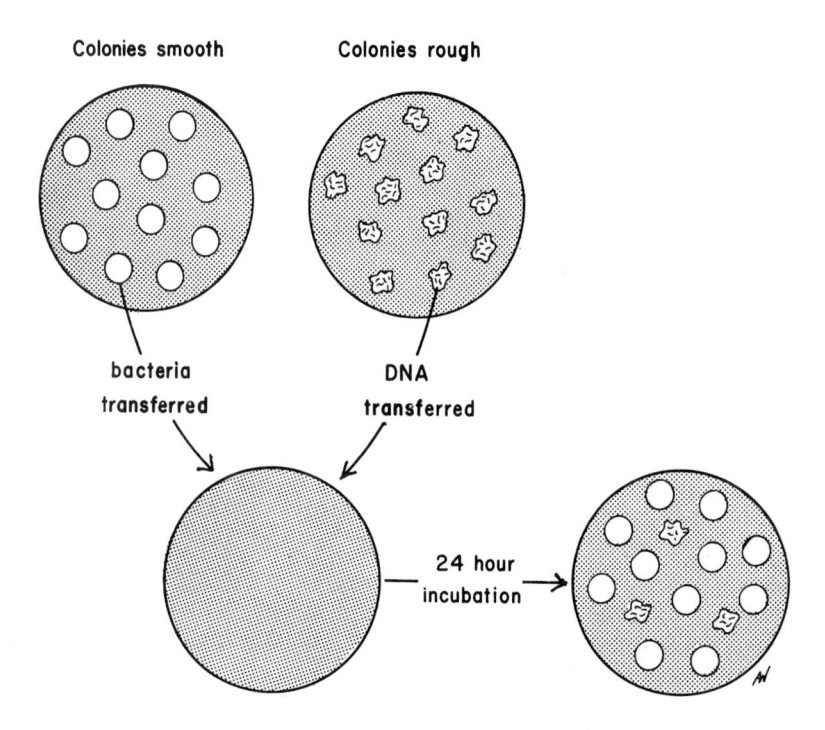

Fig. 14.1. Pneumococcus bacteria from a smooth colony are placed in a culture dish along with DNA extracted from bacteria which then form a rough colony. After an incubation period the colonies which grow are mostly smooth, but some are rough. This indicates that there has been a transference of a genetic quality from the rough to the smooth bacteria through the medium of DNA. Such transferrence is known as transformation.

One of the most striking bits of evidence comes from studies of a type of virus, the *bacteriophage,* which invades bacteria. There is one type which infects the bacterium *Escherichia coli,* a bacterium which is found in great abundance in the human large intestine. This phage (shortened form of bacteriophage) is distinguished by a polygonial-shaped body and protruding "snout." Like all viruses, it cannot feed or carry on growth and reproduction outside of the cell. Infection is accomplished when the "snout" pierces the outer membrane of the bacterium; however, the entire virus particle does not enter. Only the DNA is emptied into the interior of the bacterium with the protein coat remaining outside. Evidence of this procedure has been revealed in studies under the electron microscope and in radioactive isotope findings.

Radioactive sulfur can be incorporated into the protein and radioactive phosphorus into the DNA. There is no sulfur in the DNA and no phosphorus in the protein, so they are employed as convenient markers of these materials. These two isotopes can be distinguished by suitable detection instruments because of (1) the differences in the energy of their beta particles emitted and (2) their dissimilar half lives. When such labeled phages are placed together with bacteria, infection takes place within minutes. Thereafter, the empty coats can be washed from the outside and tested. Virtually all of the radioactive sulfur is found in these shells, and almost all of the radioactive phosphorus is found within the bacterial cells. No more than 2 or 3 per cent of the protein has entered the cell along with the DNA.

There is no visible evidence of the infection for a time, but, apparently, the DNA of the virus attracts the DNA components of the bacterium and there occurs a replication of the virus genes. Thereafter, a new protein coat is formed around each set of genes and as many as several hundred new virus particles are observed within the cell. Typically, this process is followed by a lysis (breakdown) of the membrane of the cell. The virus particles are thus free and may then infect other bacteria. The entire process, from infection to release, lasts no more than 15 to 30 minutes under favorable growing conditions.

The DNA thus forms the continuing link from one virus generation to another and as the virus exhibits detectable genetic variations we obtain this further evidence of DNA as the genetic material.

Evidence from Bacterial Transduction. A remarkable form of bacterial transformation is sometimes induced via the transport of bacterial DNA by bacteriophages. A phage may actually transport genes from one bacterium to another. Perhaps a fragment of a chromosome that escaped breakdown and conversion into phage DNA is included in the protein coat formed around the phage and in this manner is transferred when another bacterium becomes infected. This operation would be of little consequence if every invaded bacterium were destroyed, for there would then be no opportunity for expression of the transduced genes. Fortunately

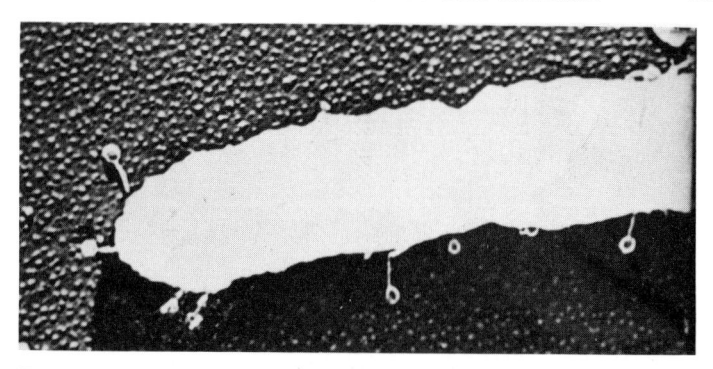

Fig. 14.2. Infection of the bacterium Escherichia coli *by a bacteriophage. This electron photomicrograph shows the greatly enlarged bacterium to which are attached the much smaller virus particles. It is through this attachment that the virus DNA is emptied into the bacterium. (Courtesy T. F. Anderson, E. Wollman, and F. Jacobs.)*

for genetic investigations, this event does not always occur. A few strains of the phage enter and establish a commensal relationship; in this way the bacteria are able to grow and reproduce. Such cells are said to be lysogenic. Viruses which have the ability to establish this type of relationship are said to be **temperate phages,** as contrasted with the **virulent phages** which destroy all cells invaded. The temperate phages on the other hand destroy about one half of the cells invaded.

The genes which have been observed to perform this method of transfer include those which carry the ability to ferment sugars; the ability to synthesize amino acids; the flagella form; and the antigenic properties. Since the virus transfers DNA almost solely from one cell to another this is cited as evidence in favor of the theory that DNA comprises the essential genetic material.

Evidence from Higher Forms of Life. Quantitative chemical measurements reveal that the amount of DNA doubles with each cell cycle from one division to another. Radioisotopes show that there is no turnover of DNA in the cell. Proteins are constantly being built up and torn down in the cell metabolism. Isotopes in amino acids reveal that they are incorporated into cells and form proteins; later, however, the proteins are broken down and the isotopes show up in the urine. This is not true of the DNA. Isotope-tagged precursors of DNA will be incorporated into the

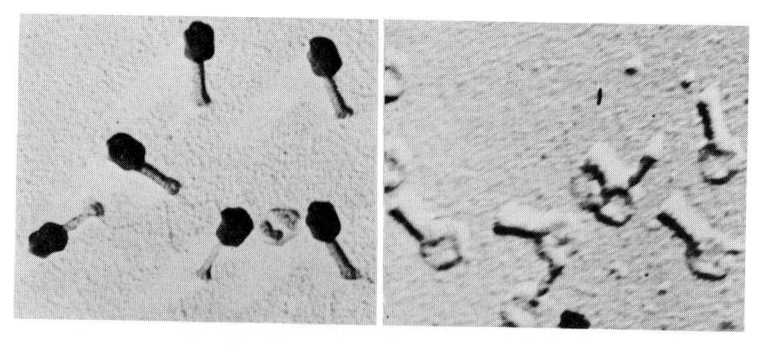

Fig. 14.3. Bacteriophage particles. The photograph on the left shows the normal appearance under the electron microscope. Those shown on the right have received osmotic shock and only the protein overcoats remain—the DNA has escaped. (From R. M. Herriott, Chemical Basis of Heredity, Johns Hopkins Press.)

nucleus of a cell, but they never leave the cell under normal circumstances. When the cell divides, some of the tagged DNA moves to each of the daughter cells; thus, the total amount per cell decreases in actively dividing tissue, whereas the amount of radioactive DNA remains the same for the entire tissue. This condition is to be expected if the genes are as stable as they are supposed.

There is some evidence of the possibility of transformation in higher animals. Jacques Benoit of the Collège de France extracted DNA from the testes of ducks of the Khaki Campbell breed and injected it into ducklings of the Pekin breed. As they matured, the ducklings tended to develop characteristics of the Khaki Campbells.

THE COMPONENTS OF DNA

As the evidence has accumulated in favor of DNA as the gene substance, attention has turned to the chemical make-up of this substance and how it is able to achieve the diversity characteristic of the genes.

Chemicals of which DNA is Composed. DNA has a very high molecular weight, numbering in the millions, yet it is composed of a combination of several comparatively simple organic molecules. There is a pentose sugar—sugar with five carbon atoms, which is known as ***deoxyribose sugar.*** It is combined with an inorganic phosphate, is phosphorulated, and forms *deoxyribose phosphate*. The chemical structure of this is shown below.

$$\text{HO—CH—CH}_2\text{—CH—CH—CH}_2\text{—O—PO}_3\text{H}_2$$

with an O bridge from the first CH to the fourth CH, and OH on the third CH.

The *ribose phosphate* found in the RNA characteristic of the genic material of a few of the viruses, is the same, with the exception of an extra hydroxyl (OH) attached to the second carbon atom.

In addition, there is a group of four organic ring-compound bases that enter into the DNA molecule. These compounds are of two kinds, *purines* and *pyrimidines*. The purines are *adenine* and *guanine*. The pyrimidines are *thymine* and *cytosine*. Their chemical structure is shown below.

PURINES   PYRIMIDINES

Adenine   Thymine

Guanine   Cytosine

The Watson-Crick Theory. Much research has gone into the problem of how the building blocks of DNA are arranged in genes. The well-founded theory proposed by Watson and Crick has received wide attention.

According to this theory the DNA molecule consists of a long double helix—two strands twisted about one another like two strings twisted together. These strands consist of the sugar molecules connected to one another by their phosphate bonds. To each sugar molecule is attached an inwardly projecting purine or pyrimidine. The two strands are held together by the hydrogen bonds connecting these bases (see Fig. 14.4.).

Chemical analysis from different organisms reveals that the proportions of the four bases may vary widely—here must lie the basis of gene diversity. However, it has been shown that (1) the number of adenine groups is always the same as the thymine groups, and (2) the guanine groups are equal to the cytosine groups. This is explained by assuming that (1) the strings are inverted in relation to one another and (2) a projecting adenine always joins to a projecting thymine from the partner helix. Guanine would always be attached to the projecting cytosine.

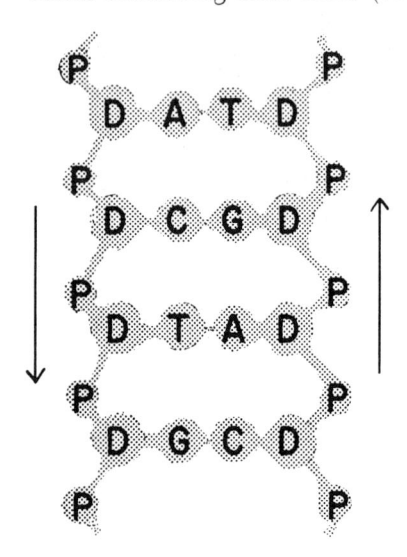

Fig. 14.4. Portion of the DNA molecule according to the Watson-Crick theory. D, deoxyribose sugar; P, phosphate bonds; A, adenine; C, cytosine; T, thymine; G, guanine. The two strings are connected at the center by the weak hydrogen bond.

Hence, the diversity which allows for the unlimited number of stereoisomeres of DNA depends upon the relative number and position of these purine-pyrimidine projections. At first glance, this theory does not seem to allow for a variation sufficient to account for all the genes of the many living organisms on the earth; however, the number of the bases in a molecule of DNA is enormous. It is estimated that the bases range from 3,000 to 30,000 or more units with a molecular weight up to 8,000,000. The various permutations of such a large group is almost infinite. It might be compared to forming words of 3,000 to 30,000 letters each from a four-letter alphabet, keeping in mind that the change of position or substitution of one letter in the entire group would make a different word.

A point in favor of this theory is that it provides a possible explanation for one of the great biological questions—how do genes duplicate themselves with such great exactitude? According to Watson and Crick, this can occur if the two helices break at their weak hydrogen bond connections between the bases, thereafter unwinding to yield single helices. Then each base could attract unto itself the proper partner—adenine attracting thymine, etc. Finally, the phosphorulated sugar is added and each helix has produced a duplicate of the helix which was separated from it.

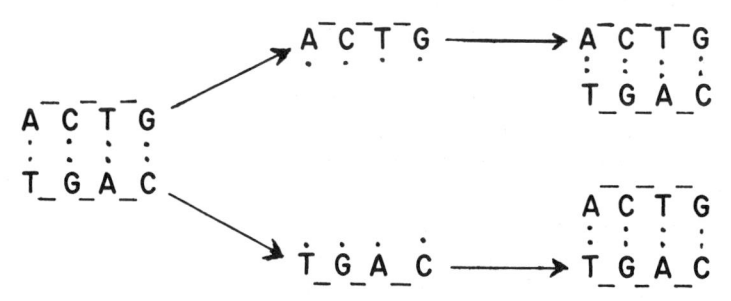

Fig. 14.5. *Gene duplication according to Watson Crick theory. This theory assumes that the double spiral which makes up the gene becomes separated into two single spirals and each attracts unto itself the chemical constituents which were removed when the spirals separated.*

THE GENE UNIT

The original concept of the genes as a series of tiny, independent, bead-like bodies linearly arranged on a gene string has had to be modified in the light of these recent discoveries. One theory (by Goldschmidt) even goes so far as to propose that genes do not exist as separate entities, but that the chromosome is a continuous chain of genetic material (a huge macromolecule) and what we call genes are merely areas of the chain which govern specific reactions. The Goldschmidt theory is supported by the fact that genes have a position effect in many instances and react differently when translocated from one part of a chromosome to another. According to this theory, gene mutations are actually small deletions or rearrangements within the chromosome and not actual changes in the physical structure of the genetic material at all. If this is true, it is difficult to see how all of the many varieties of genes could have arisen in the evolutionary process; hence, we might need to retain our concept of the gene as a unit, but with some modification of the older theories in this regard.

The Fine Structure of a Gene. The gene, traditionally, has been accepted as the smallest unit which can be recognized for recombination, for mutation, and for function. The discovery of crossing over within functional units and the establishment of pseudoalleles, however, has shown that the aforementioned can exist as three distinct units. The following series of terms to describe them has been proposed.

RECON. This is the smallest unit which can undergo recombination in crossing over. Whenever it is stated that no crossing over takes place within one portion of the chromosome, it would be more accurate to say that no crossing over has been noted in the particular number of individuals observed. Frequently, when very large numbers are employed, crossing over will be found where none had been found before. Seymour Benzer's work on this subject indicates that a recon consists of not more than two pairs of nucleotides. (A *nucleotide* is the sugar molecule with a phosphate bond and the attached base.)

MUTON. This is the smallest unit which can mutate. A change in a single nucleotide of a gene results in a mutation.

CISTRON. This is the functional unit—the part of the chromosome which governs some specific body characteristic. It is enormous in size when compared to the recon and muton. The average number of nucleotides in a cistron appears to be about 1500.

These new concepts enable us to answer many of the puzzling questions concerning gene structure. Pseudoalleles could arise through recombinations of recons within cistrons, and multiple alleles could be changes of a number of mutons within a cistron.

CYTOPLASMIC INHERITANCE

When investigations of the pattern of heredity were first begun, it was assumed that the units of heredity were located in the cytoplasm. As cytological studies continued, however, it became apparent that the genes governing inheritance were located within the nucleus, and the idea of cytoplasmic inheritance was abandoned. In more recent times, however, discoveries have been made which show that there are a few cases where bodies within the cytoplasm do transmit heritable characteristics. DNA, which was formerly thought to be found only in chromosomes, has been found to exist in small quantities in such cytoplasmic bodies as **mitochondria, plant plastids,** and **kinetoplasts** (small bodies which form the cilia of certain protozoa). These bodies can have a duplicating mechanism of their own, independent of nuclear genes. Let us now examine some of the results of research which show that there is some degree of cytoplasmic inheritance.

Plastids in Plants. In the cytoplasm of the cells of most plants there are bodies known as plastids which govern certain physiological reactions in the cells. The best-known of these are the chloroplasts, which are concerned with food manufacture. These plastids have some properties which are similar to nuclear genes. They undergo an autocatalytic duplication similar to that of nuclear genes and they can undergo mutation which may be increased in rate by irradiation. In addition, they influence specific characteristics in a manner similar to nuclear genes.

The four-o'clock, *Mirabilis jalapa*, will serve to illustrate. This plant, like many others, has variegated forms in which there are color areas of pale green or white as well as areas of the normal green. Sometimes the variegation includes small areas on the leaves and sometimes there may be entire branches of the plant that are pale green or white. Cross-pollinations can be made of the flowers on the different branches and it can be demonstrated that transmission of the trait depends solely on the ovule (female gamete). For example, an ovule in a flower located on a pale green branch will always yield pale green offspring no matter what the source of the pollen. Likewise, an ovule from a green branch will yield green offspring even though the pollen stems from a white or pale green branch. An ovule from a branch which is variegated will yield offspring of all three types—green, pale green, and variegated—no matter what the source of the pollen.

This is a case of maternal transmission of a characteristic; it is correlated with the fact that the ovule transmits cytoplasm, whereas little, if any, cytoplasm seeps through the male generative nucleus which fertilizes the ovule. The cytoplasm of the ovules contain tiny plastid primordia which divide again and again as the cells divide, and when the leaves are formed these primordia develop into chloroplasts of the same type as those of the previous generation. This process is similar to inheritance through nuclear genes; it is known that plastids are also dependent upon nuclear genes (the many cases of albinism in plants attest to the presence of such genes).

Carbon Dioxide Sensitivity in Drosophila. Carbon dioxide may be used as an anesthetic for many insects and related forms of life. Within a short time after exposure to the gas, the organisms cease activities. It was discovered that a certain race of *Drosophila* was more sensitive than others to this gas. They became anesthetized in a much shorter time and would die if exposed to a high concentration. When females from this race are crossed to males of a normal race, the offspring are all sensitives. The reciprocal cross, however, yields no sensitives. Each of the chromosomes of the sensitive race has been replaced by those from a non-sensitive race through outcrossing, yet the results have not been altered.

This first appeared to be a case of cytoplasmic inheritance, but it was later found that sensitivity could be induced in non-sensitive strains by injecting extracts of the sensitives. This event led to the conclusion that the characteristic was probably due to a virus which was located in the cells of the sensitive strain and which was transmitted through the cytoplasm of the eggs down through the generations. Therefore, these could not be classified as examples of true cytoplasmic inheritance.

Kappa Bodies in Paramecium. A very interesting case of apparent cytoplasmic inheritance in *Paramecium* was found by T. M. Sonneborn of Indiana University. He observed that some paramecia (killers) produce a substance known as *paramecin* which can kill other paramecia (sensitives) that do not produce this substance. The killers were found to contain cytoplasmic bodies, *kappa*, which produce the paramecin. Extensive studies of the conjugation of these one-celled organisms revealed that the killer trait was transmitted through the cytoplasm. As in the case of *Drosophila*, however, it was found that sensitives could be converted into killers by transmission of kappa particles. When all evidence is considered, it appears that the aforementioned is a case of invasion by some virus-like bodies and not a case of true plasmagenic inheritance.

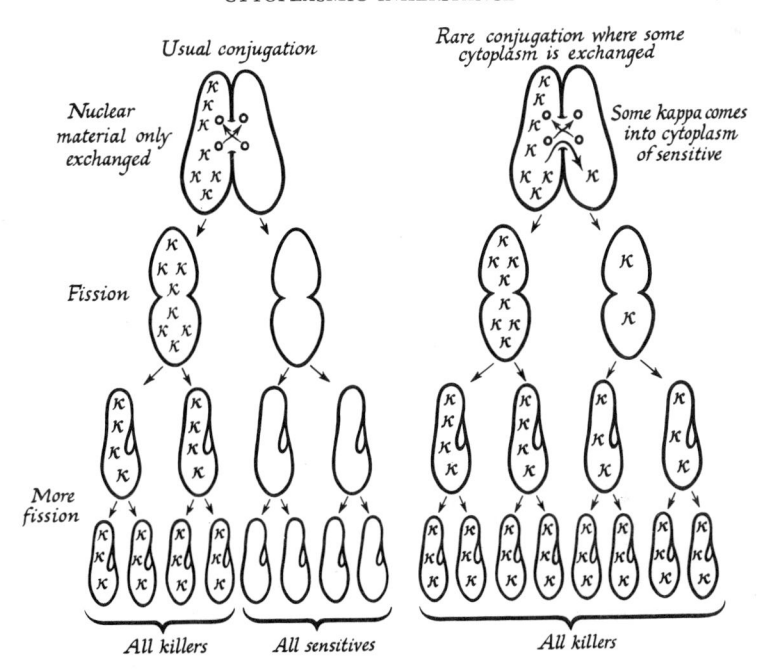

Fig. 14.6. Evidence of transmission of gene-like bodies in the cytoplasm of Paramecium. In rare cases of conjugation between killers and sensitives some of the cytoplasm is transferred into the sensitive protozoan. The kappa bodies that thus gain entrance multiply and transform the sensitive into a killer, and its offspring are likewise killers. (From Winchester, Genetics, Houghton Mifflin.)

Cytoplasmic Inheritance in Haploid Organisms. The best evidence for cytoplasmic inheritance has come from studies of haploid organisms. One strain of yeast is red and another is white, but when the red strain is stimulated to grow very rapidly, some white cells are formed and these white cells have only white descendants. This observation can be explained by an assumption that some factor for red pigment is located in the cytoplasm. Yeast reproduce by budding, and when the cells are growing very rapidly the factors for red may not reproduce fast enough to insure that at least one factor for red is included in each bud which is formed. Those buds not receiving the factor grow into white cells.

In the mold, *Neurospora*, a strain has been found which is called *poky* because its filaments grow so slowly. This trait has been found to be transmitted by a cytoplasmic factor which alters the respiratory enzymes causing them to function improperly.

Investigations by Ruth Sager on the green alga, *Chlamydomonas*, also show clear-cut cytoplasmic inheritance. This one-celled organism is haploid, but at times two cells of opposite sex unite to form a diploid zygote. Meiosis takes place later, and four haploid cells are produced from the zygote. Two of these haploid cells will be female and two will be male because of a nuclear gene which determines sex. When sexual union occurs, both male and female cells contribute cytoplasm, but inheritance of cytoplasmic factors takes place mainly through the female cell. Perhaps the male cytoplasmic factors are cast off during divisions of the zygote in meiosis. Bacteria are known to throw off extra genes when they become partially diploid through conjugation, so it is possible that a similar occurrence happens here. Such characteristics as streptomycin resistance have a pattern of inheritance which indicates that they are transmitted through cytoplasmic factors.

In this section we have shown that there are definitely some cases of inheritance through the cytoplasm. Most characteristics certainly are controlled by genes on the chromosomes in the nucleus, but the replication of such bodies as chloroplasts and mitochondria are better geared to the environmental needs of the organism and these have independent dividing mechanisms. The cytoplasm also carries some factors of inheritance of characteristics not related to the chloroplasts and mitochondria.

15: GENE ACTION

With the great strides that have been made in the techniques of biochemistry, it has become possible to investigate the methods of gene action which result in various phenotypic effects. Until recent times geneticists knew only that if a person was homozygous for a particular recessive gene he was an albino—what went on in the body to cause this condition was a mystery. Today, however, we have considerable information about the alterations of body chemistry which these genes create and we now know the reason why albinism results. Such information promises to have more than theoretical interest. Knowledge of the exact chemical pathways which have been interrupted or altered by specific genes can lead to a possible means of correction of inherited abnormalities. For example, a certain type of inherited idiocy has been found to result from genic alteration of the cellular metabolism of a common food constituent. With this knowledge as a basis, it has been found that the idiocy can be prevented by the simple procedure of withholding this food constituent from the diet. It is the aim of the present chapter to summarize such discoveries concerning the method of gene action.

HUMAN HEMOGLOBIN

Hemoglobin is the oxygen-carrying part of the red blood cells; it is considered a vital blood constituent because a constant oxygen supply is necessary for all active body cells. Recent investigations of this substance have given geneticists useful information about the ways that genes can influence body chemistry.

Sickle-cell Anemia. Among peoples native to northern Africa and the Mediterranian countries of southern Europe there frequently develops an inherited blood abnormality known as **sickle-cell anemia.** This is a severe disorder and afflicted persons usually die before reproduction, but there is a much more common condition, known as the *sickle-cell trait,* which causes little or no interference with normal activities. About 8.0 per cent of American Negroes have the sickle-cell trait, but only about 0.2 per cent develop the sickle-cell anemia. Both conditions are practically unknown among Americans of northern European descent.

Fig. 15.1. Red blood cells from person with sickle-cell anemia. The hemoglobin of the cells tends to crystalize when there is a low oxygen concentration and this crystalization causes the cells to form the odd shapes shown here. (Courtesy James V. Neel, Heredity Clinic, University of Michigan.)

A microscopic study of a blood smear taken from a person with sickle-cell anemia shows that the red blood cells are not the small biconcave discs characteristic of normal blood; instead they are formed into odd crescent or sickle shapes. Such cells are poor conductors of oxygen—hence the anemia. The blood from a person with the sickle-cell trait has cells of normal appearance if the smear is made as usual on a slide in the air. However, if such cells are held under conditions of very low oxygen concentration for a short time before the smear is made, the cells will express the sickle shape.

The genetic basis of these conditions was recognized as early as 1923, but it was not until 1949 that the now generally accepted pattern of inheritance was proposed by J. V. Neel of the University of Michigan. This theory assumes that the gene involved is intermediate in nature—the sickle-cell trait is the heterozygous expression of the gene, whereas the anemia is a result of homozygosity.

Linus Pauling was one of the first to show that there is a chemical difference between the normal and sickle-cell hemoglobin. He employed the principle of electrophoresis to separate the two. When *hemoglobin A* (normal) was placed on a paper soaked with a buffer solution and exposed to an electrical field, the hemoglobin tended to move toward the positive pole. *Hemoglobin S* (taken from blood cells of a person with anemia), however, tended to move toward the opposite pole. Hemoglobin taken from a heterozygous person separated with some movement toward one pole and some toward the other, thus suggesting that both A and S were present. This supposition was supported by the fact that a mixture of hemoglobins (from a normal person and from an individual with the anemia) likewise segregated out, forming a bimodal distribution.

V. M. Ingram, of Cambridge University, carried out a detailed chemical analysis of these two hemoglobins. The hemoglobin molecule is relatively enormous; it is composed of two symmetrical half-molecules, each consisting of about 300 amino acids. This unit is too large for an effective analysis. Ingram, therefore, broke down the molecules by using the enzyme trypsin, which causes the chain of amino acids to break at points where lysine and arginine are found. By this technique he obtained 28 fragments, each containing a small group of amino acids known as peptides. Through the use of electrophoresis followed by paper chromatography, he found that S differed from A in only one peptide of the group. Both the S and A types were composed of the same kinds of amino acids, but in the S hemoglobin there were two valines and one glutamic acid, whereas the A hemoglobin contained one valine and two glutamic acids. Thus, there was a substitution of valine for glutamic acid at one position (illustrated in Fig. 15.2.). This single gene-induced change in one of the 300 amino acids of each half-molecule is sufficient to account for the sickling effect. Since glutamic acid has a negative charge and valine has no charge, this factor would explain the difference in electrophoretic properties.

The sickling of the hemoglobin is due to the great difference in the solubility of the two—S is only about 2 per cent as soluble as A and it begins to crystalize when there is a lowering of the

HEMOGLOBIN A

histidine — valine — leucine — leucine — threonine — proline — glutamic acid — glutamic acid — lysine

HEMOGLOBIN S

h — v — l — l — t — p — valine — g — l

HEMOGLOBIN C

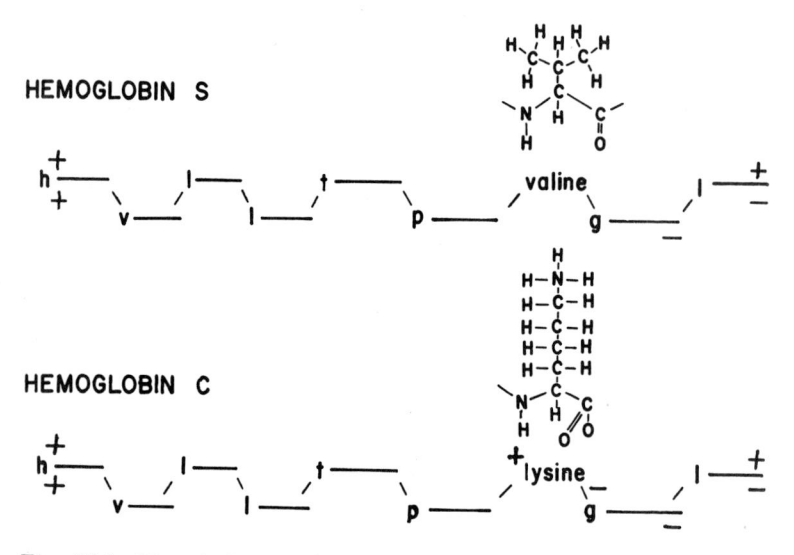

h — v — l — l — t — p — lysine — g — l

Fig. 15.2. *Chemical differences in three kinds of hemoglobins. This shows only one peptide out of 28 which make up a half-molecule of hemoglobin, but the only difference between the hemoglobins lies in this one peptide. The structural chemical formula for the one amino acid which varies is shown.*

oxygen concentration. The crystalization causes the distortion of the red blood cells (as seen in the blood smear) and, in addition, increases the fragility of the cells so that they tend to break more easily. The sickle-cell formation is most likely to take place in the capillaries where the hemoglobin gives up its oxygen in the reduced form which crystalizes easily. These odd-shaped cells tend to impede the circulation and this causes still further reduction in oxygen concentration with further sickling in the manner of a vicious cycle.

Heterozygous persons have only about 24 to 45 per cent hemoglobin S; this quantity is not sufficient to cause sickling in normal concentrations of oxygen in the blood, but under conditions of reduced blood oxygen (which could occur as a result of either poor circulation or removal to a high altitude) a certain amount of sickling will take place. Homozygous persons have been generally assumed to have 100 per cent hemoglobin S, but some cases have been recorded where the percentage was as low as 70 per cent of the total; the latter condition can affect the severity of the anemia. Studies of the blood of such persons shows that the extra 30 per cent is not hemoglobin A but is instead a fetal hemoglobin (F). The last named is a type of hemoglobin produced in the fetus, normally persisting until about six months of age, but sometimes (in small amounts) throughout life. The larger quantities found in some cases of sickle-cell anemia are probably due to a reactivation of the fetal hematopoietic mechanism as a response to a chronic anoxia arising in the body tissues.

Malarial Resistance and the Sickle-cell Trait. The geographical distribution of the S hemoglobin in world populations lead some investigators to seek out correlative factors. Could there be some difference in the environments which caused a greater number of the genes in some regions than in others? Indeed there could; it was found that, in general, the prevalence of the genes for S hemoglobin was widespread in those regions of Africa and Europe where malaria was most abundant. Further studies showed that those individuals expressing the sickle-cell trait were at the same time resistant to an infection of the falciparum malarial organism. Thus, while there is a constant selection against the homozygote because of the low viability and fertility, there is a very strong selection for the heterozygote; in this manner a balance is established. In northern Europe, where malaria has been virtually nonexistent, whenever a mutation appears the selection has only been against the S hemoglobin.

Other Hemoglobins. As in the case of the blood antigens, the discovery of the hemoglobin variations has led to further research whereby other forms have been discovered. Hemoglobin C was observed in an American Negro and since then, records reveal that almost 2 per cent of American Negroes express this trait. In Africa its distribution is variable, having no apparent correlation with the distribution of the S hemoglobin. Hemoglobin C is formed by an allele of the gene for S and is caused by a substitution of lysine for the glutamic acid in the very same peptide which is varied in the S hemoglobin. Homozygosity causes an anemia, but it is much milder than sickle-cell anemia. An interesting fact is that heterozygous C and S individuals have sickle-cell anemia.

The discovery of hemoglobin C was followed in rapid succession by the discovery of D, E, G, H, I, J, K, L, N, O, P, and Q. At least two loci appear to be involved.

Hemolytic Reactions to Drugs. During World War II, when the quinine supply was cut at a time when needed most, an intensive search was made for other antimalarial drugs. One of those discovered, known as primaquine, proved to be an effective agent in the prevention and treatment of malaria, but its use was limited when it was found that in some cases a severe hemolytic reaction followed administration of the drug. This reaction proved to be an inherited condition which apparently is caused by the loss or inactivation of an enzyme in the erythrocytes. This enzyme, glucose-6-phosphate dehydrogenase, seems to be able to break down harmful intermediary products of the primaquine, but in its absence the hemolytic reaction occurs. B. Childs, of the Johns Hopkins University, has demonstrated that the sensitivity to the drug is inherited as an intermediate sex-linked gene.

The same enzymatic deficiency is the apparent cause of the well-known hemolytic reactions to some sulfa compounds and other drugs of a similar nature. This discovery turns our attention to the relationship between genes and enzymes.

GENES AND ENZYMES

We are familiar with the enzymes in the digestive system of man and other animals. The enzymes cause chemical changes in digested food, thus converting it into a form which the body can absorb. There are also cellular enzymes which bring about chemical changes within the cells. It appears that each enzyme accomplishes only one chemical change and, considering the numerous chemical reactions which occur within any given cell, the number of such enzymes must indeed be very great. We know that many genes accomplish their effects on the body by means of

GENES AND ENZYMES

the enzymes which they produce. Each gene of this type apparently produces only one enzyme and this contributes somewhat to the total chemical reactions in the cells. However, this small contribution can be very important—the failure of even one chemical reaction in a series can have far-reaching effects.

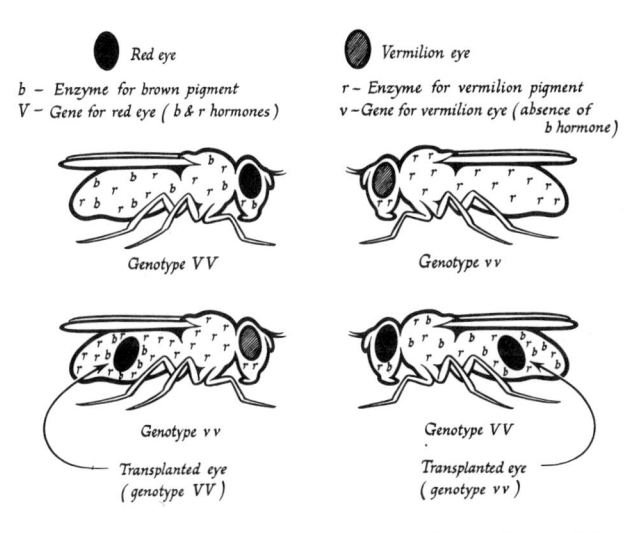

Fig. 15.3. Enzymes influencing eye color in Drosophila. *These diagrams show how the enzymes for brown and vermillion pigment influence the development of an eye which has been transplanted into the abdomen of a fly with a different genotype. (From Winchester,* Genetics, *Houghton Mifflin.)*

Drosophila Eye Color Pigments. The eye color of the wild-type *Drosophila* is a particular shade of red. Chemical analysis shows it is caused by a blend of two pigments, one **vermillion** (orange-red), the other, **brown**. When a fly is homozygous for the gene, *v*, the brown color does not appear and the eyes are vermillion. According to the enzyme theory, the condition could be caused by the failure of the particular gene to produce an enzyme necessary for the production of the brown pigment. Dr. George Beadle, of the California Institute of Technology, employed an ingenious technique to test this theory. He removed tissue from the larva of *Drosophila* at the site where an eye would form and transplanted it into the body cavity of another larva. Oddly enough the transplanted tissue continued to grow and actually produced an eye within the body cavity of the host. When eye tissue from a wild-type larva was placed in the body of a homozygous vermillion larva, an adult emerged with vermillion eyes on its head and a red eye in its abdomen. It would seem that the transplanted cells produced the necessary enzymes for the formation of both pigments. When the procedure was reversed and tissue from homozygous vermillion was transplanted into a wild-type larva, the eyes on the head were red and so was the eye in the abdomen. This can be explained by assuming that the cells surrounding the transplanted tissue produced the enzyme necessary for the formation of brown pigment and that this diffused into the transplant.

There are many other genes which affect eye color in *Drosophila*. These appear to affect the quantity of one or both pigments, the acidity of the cells which will alter the color, and oxidation or reduction of the brown pigment. All of these conditions appear to be based on enzymatic reactions.

ENZYMES IN NEUROSPORA

The pink bread mold, *Neurospora*, has proved extremely valuable in studies of genes and enzymes and is one of the most widely used organisms in such genetic investigations today. Its main vegetative body is haploid and, therefore, no complicated test crosses are necessary—all genes carried will be expressed. Furthermore, it can be propagated asexually so that a strain can be continued as long as needed without genetic change. In addition, it engages in a sexual union of gametes with all the advantages of sexual reproduction.

The main body of the mold consists of a mass of interwoven threads. They grow by extending themselves into the medium from which the mold obtains its nutrients. When mature the mold produces gametes, fertilization occurs, and a diploid zygote is formed. This process does not produce diploid tissue, however, because the first two divisions are the meiotic divisions wherein four haploid cells result. A third division follows which is mitotic in nature wherein there are eight haploid spores, each of which can form a new plant body. The spores may be genetically different. If the parent molds were different with respect to a particular gene, four of the spores will carry the gene as found in one parent and four will contain the gene as found in the other. It is possible to pick up these spores one at a time by using a dissecting microscope; from a single spore a pure culture can be started.

Tryptophan Synthesis. *Neurospora* is able to live on culture media with the barest of nutrients. Only a few simple salts, sugar, ammonia, and biotin (a B vitamin) are necessary for the growth of the normal, or wild-type, form of this mold. This factor does not mean that the *Neurospora* does not need the other vitamins and the amino acids such as are generally required for other forms of life. *Neurospora* can synthesize its own vitamins and amino acids from the simple substances mentioned. It does this through enzymes.

This process can be illustrated by the method of synthesis of the amino acid, *tryptophan*, from *indol* and *serine*. A gene produces an enzyme, *tryptophan synthetase*, which causes the indol and serine to combine with the extraction of a molecule of water; tryptophan results. A mutant form of the gene apparently does not produce this enzyme. It is noted that the mold bearing this gene cannot grow on a medium without tryptophan even though serine and indol are present.

One might assume that these results could be explained by a small deletion or inactivation of the gene which produces the enzyme. Recent discoveries reveal, however, that it is probably a change in the chemical nature of the enzyme rather than its elimination. Suskind and Jordan, of the Johns Hopkins University, discovered that the tryptophan-requiring mutants contained an enzyme closely related to the tryptophan synthetase. It can act upon indol in the presence of triose phosphate, hence forming indole glycerol phosphate. This enzyme is not found in the wild-type *Neurospora*, and its similarity in chemical nature and function to tryptophan synthetase proves that the mutation has resulted only in a slight alteration of the enzyme.

Reverse mutations of the mutant mold back to the wild type occur in rare instances, minimizing the theory that the deletion or inactivation of the gene causes the original loss of ability to synthesize tryptophan. The reverse mutations are easy to detect. Millions of spores from a strain which does not produce the enzyme can be placed on a medium containing serine and indol, but not tryptophan. If there are any reverse mutants among these millions, they will grow and the new strain can be isolated from such growths.

Enzymes in Series. These examples of enzymes in *Drosophila* and *Neurospora* may give a deceptively simple picture of the gene-enzyme relationship. It is really much more complex than the examples might lead one to believe. There is usually not one, but a long series of enzymes and reactions in most of the processes of

TRYPTOPHAN SYNTHESIS IN NEUROSPORA

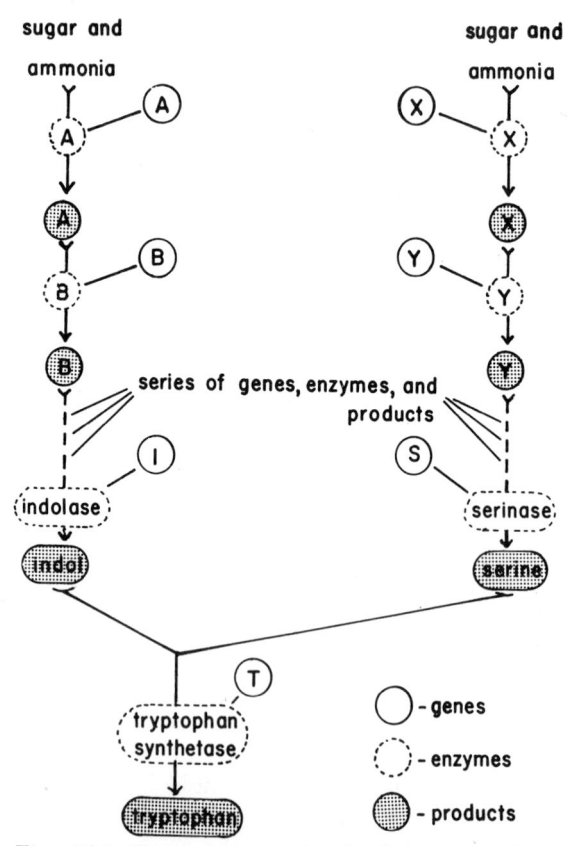

Fig. 15.4. *Enzyme series involved in tryptophan production in* Neurospora. *It is because of extensive enzyme systems such as this that this mold can exist on a medium containing a minimum of ingredients.*

synthesis or degradation that go on within the cell. Indol and serine are the final products which are formed just before tryptophan production. *Neurospora* developing from the mineral medium of salts, sugar, ammonia, and biotin has no serine or indol to act upon, but it can produce these two substances from sugar and ammonia by a series of reactions (see Fig. 15.4.). A mutation of one gene along the chain leading to serine would eventually result in the prevention of tryptophan production. The same is true for the chain leading to indol.

A similar chain of enzymes is necessary for the synthesis of each of the other amino acids—for example, leucine, valine, arginine, lysine, etc. In addition, there is the rather extensive group of vitamins which must be synthesized. When we add to this the other enzymes necessary for the numerous metabolic activities of the mold, we can appreciate the large number of enzymes which must be produced.

AN ENZYME SERIES IN MAN

It is much more difficult, of course, to trace the action of genes through enzymes in higher forms of life, but the techniques of biochemistry have become so advanced that it is possible to work out some rather detailed pathways. In the process of digestion, the proteins in foods are broken down into their component amino acids which can then be absorbed through the intestinal wall and thence carried through the body. The amino acids are then absorbed by the cells where they may be reassembled into proteins—or by way of a chain of reactions, governed by cellular enzymes, they may be degraded into carbon dioxide, water, and mineral waste, with a subsequent release of energy. We shall consider the degradation of one amino acid—phenylalanine—which is the best understood from a genetic viewpoint.

Phenylalanine Metabolism. This amino acid may follow one of three pathways depending upon the enzyme which acts upon it. It may go into the production of cell protein, it may be converted into another amino acid known as tyrosine, or it may be converted into phenylpyruvic acid. A dominant gene, P, is necessary for the production of the enzyme for the conversion into tyrosine. The recessive allele of this gene causes some alteration of the enzyme or blocks its action so that this conversion does not take place.

This causes an accumulation of the phenylalanine, and an abnormal amount of it is converted into phenylpyruvic acid, which accumulates in the cells in large amounts. Some of it diffuses out of the cells, resulting in a high concentration of the acid in the blood. In such concentrations it appears to poison the nervous system, for the condition is accompanied by mental derangement. Persons so afflicted are known as *phenylpyruvic idiots* or *imbeciles.* In such cases, mental derrangement can be prevented if a person is fed on a diet free of phenylalanine (this amino acid can be easily removed from protein foods by chemical extraction). The diet must begin early in life, however, because the brain damage appears to be irreversible. The acid is excreted in the urine and it is readily detected by a simple chemical test—a factor of great practical significance. It is now recommended that the urine test be made on all children who show any signs of mental deficiency. About one child out of every 25,000 born will be homozygous for the recessive gene causing this condition.

In addition, tyrosine, depending upon the enzymes which act upon it, has three pathways of conversion. One involves its conversion into *dihydroxyphenylalanine* through the action of gene A. This substance is then carried through a series of changes mitigated by other enzymes and forms the pigment *melanin* which is found in an individual's skin and hair and in the iris of his eyes. When a person is homozygous for the recessive allele, *aa*, the normal course of action is blocked and the melanine is not produced. Without this pigment the individual is *albino.*

In most cases a different gene acts upon tyrosine and converts it into *parahydroxy-phenylpyruvic acid.* This acid can also be made from phenylpyruvic acid. Another enzyme converts the parahydroxy-phenylpyruvic acid into *dihydroxy-phenylpyruvic acid.* A rare recessive gene blocks the conversion at this point, a build-up of the parahydroxy-phenylpyruvic acid results, and acid will be excreted in the urine. This consequent condition is known as *tyrosinosis.*

Still another enzyme in the series forms *homogentisic acid* which is followed by another enzyme-stimulated reaction with the result that *acetoacetic acid* is formed. The recessive gene, *h*, however, may form a block at this point so that homogentisic acid accumulates in the system and is henceforth excreted. This homogentisic acid is also known as *alkapton*, and the condition it causes

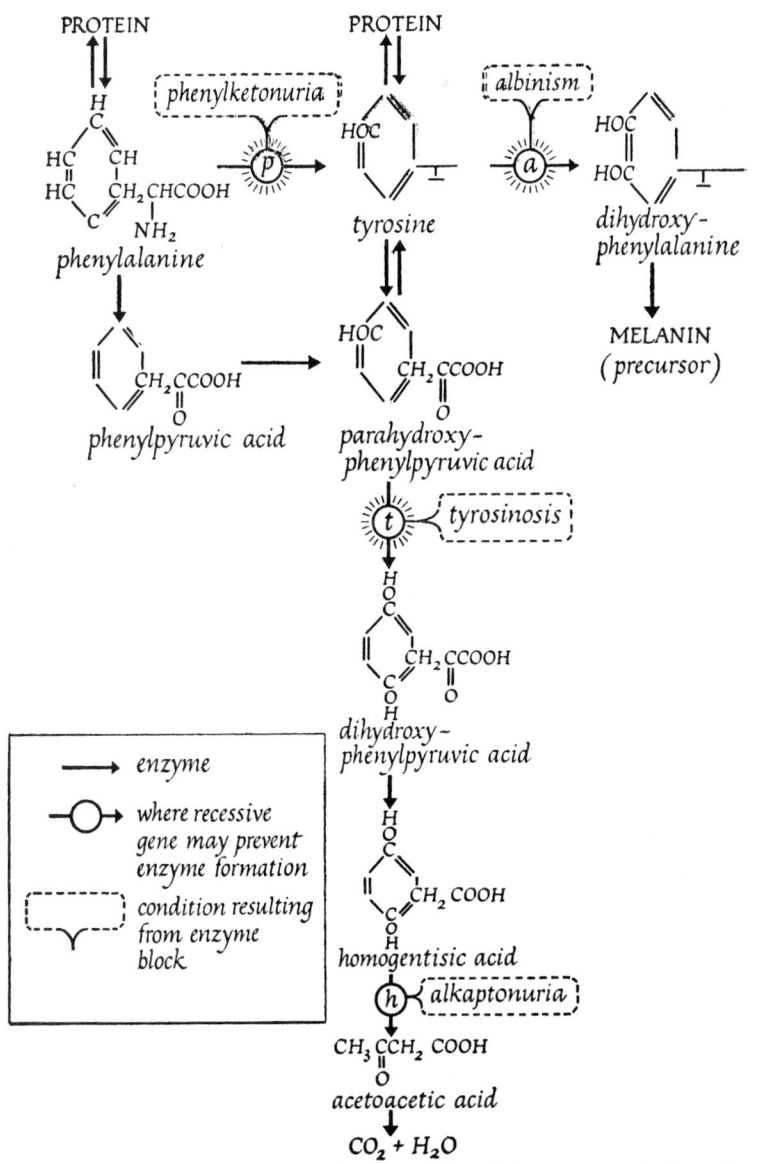

Fig. 15.5. Enzyme series involved in phenylalanine metabolism in man. The four places where breaks are known to occur in the series are indicated together with the type of abnormality which results from the break. In the structural formulae only those portions are shown which represent a change from the original molecule. (From Winchester, Genetics, Houghton Mifflin.)

is called *alkaptonuria*. This condition is easy to recognize because the urine turns dark after being exposed to the air for a short time. A mother usually first detects it in her baby because the soiled diapers turn black. The condition is not a serious affliction for most alkaptonurics, but in some cases, where the alkapton has penetrated the cartilages of the joints, there may result a painful form of arthritis. In addition, the cartilages of the ears and nose often become darkened because of the accumulation of the acid, and in a few cases the skin and the sclera (white) of the eyes are discolored. Still other enzymes carry on the chain of degradation to carbon dioxide and water.

Thus, the three known gene substitutions in the group of genes involved in phenylalanine metabolism can have distinctive phenotype effects. When we consider all of the amino acids as well as the other food substances which are influenced by enzymes, the possible number of variations is staggering to the imagination.

Enzymes and Gout. As research continues there is little doubt that many of man's diseases will be found to be caused by enzyme alterations of the cell metabolism. One of the latest findings concerns the well-known disease called gout. The disorder is characterized by the accumulation of excessive quantities of **uric acid** in the blood, with a very painful swelling of the extremities and resulting arthritis. It was believed that the disease was caused by an intake of a protein rich diet, complicated by a kidney deficiency which failed to remove uric acid from the blood in sufficient quantities to prevent its accumulation. Studies of the origin of uric acid show that it is formed by cellular enzymes from glycine. When glycine tagged with radioactive carbon 14 is administered to a person with an inherited predisposition to gout, there will be an excretion of up to four times as much of the radioactive carbon in the uric acid as there is in a normal person. Thus, it is not a deficiency of the kidneys in excreting uric acid that causes the disease but rather an excessive production of the acid by enzymes in the cells. This is yet another example of how genes exert their great influence on the body through enzymes.

THE TRANSMISSION OF GENETIC INFORMATION

We have learned that, in most cases, the information for the growth and activities of cells is contained within the genes in the nucleus of the cells, yet we also know that the seat of synthesis of protoplasm and of the cellular enzymes is located in the cytoplasm. The cytoplasm contains very small bodies, known as *ribosomes,* which are the sites of protein formation. It is at the ribosomes that the amino acids are assembled into the chains which, in turn, form proteins. How is the information within the genes transmitted to the ribosomes so that the amino acids are assembled in the proper order according to the type of cell involved?

Gene Messengers. Genes send messages out to the ribosomes in the form of messenger RNA (ribonucleic acid). RNA is very similar to DNA, but there are three important differences. First, the sugar in RNA is ribose sugar instead of deoxyribose sugar. Second, the pyrimidine, uracil, is substituted for the pyrimidine, thymine. Third, the RNA strand is single rather than double as in DNA.

The formation and transmission of *m-RNA* appears to be as follows. When the gene is to produce m-RNA, the two DNA strands separate at their weak hydrogen bonds as if the gene were going to replicate. Instead of replicating, RNA is formed along one of the two strands in complementary fashion, with the exception that uracil, rather than thymine, is placed on the RNA opposite adenine on the DNA. Ribose sugar and phosphate then is added, making a single strand of m-RNA. Since genes average about 1500 paired nucleotides in length, the messenger RNA will average about 1500 single nucleotides in length, since each m-RNA carries the message or code of a single gene. The m-RNA then separates from the gene and passes out into the cytoplasm where it contacts the ribosomes. It appears that several ribosomes may be involved in the transcription of the message from a single gene. These are known as *polyribosomes.*

The Transfer of Amino Acids. Before protein synthesis can take place the amino acids must also be carried to the ribosomes. This is accomplished by means of another type of RNA, transfer RNA, which is also called soluble RNA, abbreviated to *t-RNA* or s-RNA. This RNA is much shorter in length than m-RNA having only about 87 nucleotides, and it is twisted about itself so that it appears somewhat like a twisted hairpin. Each t-RNA molecule is coded to carry one particular amino acid to the ribosomes. Ribosomes also contain RNA (ribosome RNA) which may serve as an attractant for the other two types of RNA.

At the ribosomes the m-RNA furnishes the code and the t-RNA

brings the amino acids which are fitted together in a chain according to the code. Thus, a *polypeptide chain* of amino acids is formed. This polypeptide chain itself may be a protein molecule, or a group of two or four chains may be joined to form the protein molecule. The chains are folded and cross-linked to form a molecule more compact than would be possible with unfolded chains. The proteins may be structural proteins which make a part of the protoplasm, or they may be functional proteins which act as enzymes to mediate many cell reactions.

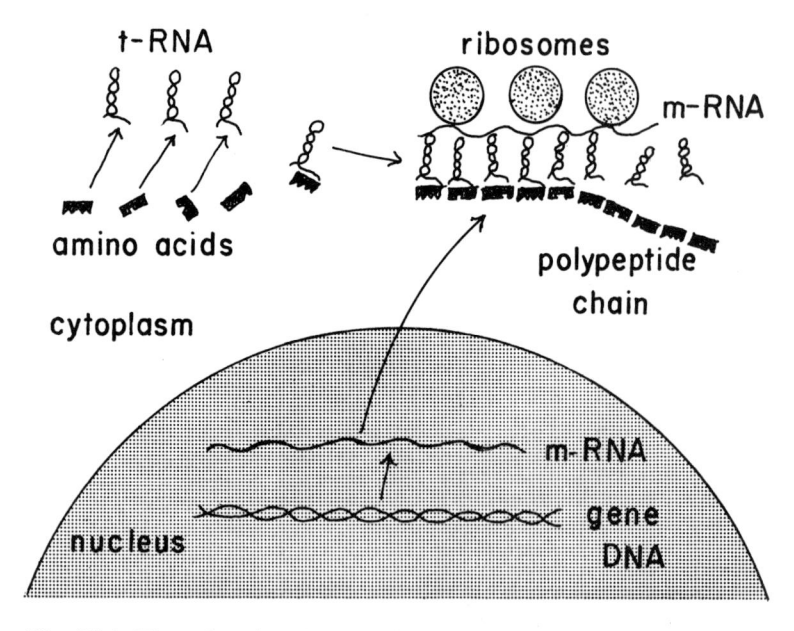

Fig. 15.6. The roles played by DNA, RNA, and the ribosomes in protein synthesis. Messenger-RNA is produced by the gene (DNA) in the nucleus and moves out to the ribosomes in the cytoplasm. Transfer-RNA picks up the free amino acids in the cytoplasm and transports them to the ribosomes where they are assembled into a polypeptide chain according to the code on the m-RNA. The polypeptide chain may be a protein, or it may combine with other such chains to make the protein molecule.

THE GENETIC CODE

Extensive research has gone into the job of trying to "crack" the genetic code. The results of this research indicate that a series of three bases in the m-RNA carries the code word, or **codon,** for one of the 20 amino acids. For instance, three uracils, UUU, is the codon for the amino acid, phenylalanine. This was determined by preparing a synthetic m-RNA made only of uracil and placing it along with mixed amino acids in a suspension of extracted ribosomes and t-RNA from bacteria. From this mixture a chain of phenylalanine was formed. When adenine was added to the synthetic m-RNA, some triplets were formed which were UUA. Then it was found that tyrosine was also found in the amino acid chain which was formed. In this manner it has been possible to establish the codon for each amino acid, although many codons are still only tentatively formulated because it is difficult to determine the exact sequence of bases in the mixed triplet codons.

Some amino acids are coded by more than one triplet codon. There are only 20 amino acids in the protein make-up of the majority of living organisms, yet there are 64 possible three-base combinations. For instance, phenylalanine is coded by UUC as well as by UUU. There are some triplets, known as nonsense codons, which code no amino acids. It is believed that nonsense codons are found on the DNA of the chromosomes at the beginning and end of each gene. Such nonsense codons may act as "punctuation marks" to indicate where the triplet sequence of RNA shall begin to be formed. This is important because an entirely different genetic code would be found on the m-RNA if it began to form its triplet codons even one base removed from the beginning of the gene.

We commonly refer to the codons as they are found on m-RNA, but you should keep in mind the fact that the codons on the gene will be complementary bases. For instance, UUU on m-RNA is formed by the sequence AAA on the DNA strand of the gene. Also, the t-RNA codon is complementary to that found on m-RNA. The t-RNA which brings phenylalanine to the ribosomes has the triplet codon AAA at the twist of the "hairpin" although the amino acid is picked up by an extension of the RNA strand out at the open end of the "hairpin."

GENE MUTATIONS

The discoveries on the nature of the genetic code have made possible an understanding of the possible method of gene mutation. If a single nucleotide pair in the DNA ladder of a gene is changed, then one amino acid of the polypeptide chain which is formed at the ribosomes will probably be changed. This change of one amino acid can have far-reaching effects on the resulting protein as we have already learned in connection with the hemoglobin molecule. A mutation would result, therefore, if a nucleotide pair were reversed in its position, if a different nucleotide pair were substituted, if a nucleotide pair were deleted, or if a nucleotide pair were added. All these events would result in different codons for one or more amino acids. Many of the changes would be lethal in effect because the changed polypeptide chain might be a part of an important enzyme and a vital cell reaction might not take place. Also, a nonsense codon might result in a shortened chain which would be ineffective. Or, most of the amino acid chain would be different if there were a loss or addition of a nucleotide because the triplet sequence for the balance of the gene would be altered in such cases. On some occasions, the altered nucleotide would not change the protein because the new triplet might happen to be an alternate codon for the same amino acid as was coded by the original triplet. The following table shows what could happen as a result of the change of one letter in the codon for alanine.

VARIATIONS POSSIBLE THROUGH SINGLE BASE SUBSTITUTION OF THE CODON FOR ALANINE

Codon for Alanine	Variations Possible
CAG alanine	CAU nonsense—no amino acid CAA sense—asparagine CAC sense—proline CCG sense—alanine (no change) CGG nonsense—no amino acid CUG sense—alanine (no change) UAG nonsense—no amino acid GAG sense—glycine AAG sense—glutamic acid

CONTROL OF GENE ACTIVITY

An important question concerning gene action still remains to be answered. Why do not all genes function all of the time? A person has genes for eyes on the bottom of his foot, yet eyes are not found on this part of the body. Probably no more than 5 to 15 per cent of the genes in human cells are producing RNA at any one time. Some controlling mechanism must exist whereby genes can be turned on and off. Work on bacteria by F. Jacob and J. Monod indicates a pattern somewhat as follows. There are groups of *structural genes* on the chromosomes which control particular cell activities. Associated with each group of structural genes there is an *operator gene* which produces a substance which stimulates the structural genes to open out and produce RNA. The unit formed by the operator gene and the structural genes it controls is known as an *operon*. Finally, there are **regulator genes** which produce substances which act to destroy the stimulatory products of the operator genes. Hence, with all these systems functioning, there would be no RNA synthesis by the genes. The regulator genes, however, are sensitive to factors from outside the cell. These factors include such things as hormones and embryonic organizers which can destroy the products of the regulator genes. The operator genes then are free to stimulate the structural genes and RNA is synthesized.

Histones and Gene Activity. Chemical analysis of chromosomes shows that DNA is closely associated with the proteins known as *histones.* Experiments show that histones inhibit gene activity. Artificial removal of histones from cells results in a greatly increased output of RNA from the genes. A hypothesis to explain the relationship of the histones to the gene complex holds that histones become associated with the products of the regulator genes and together they repress gene activity. Both substances are required. Histones are non-specific as to the genes they affect since histones do not exist in sufficient variety to have a separate histone for each operon. The product of the regulator genes, however, is specific, but it needs the histone before it can function.

The Puffing Phenomenon of Salivary Gland Chromosomes. Studies of the large salivary chromosomes of *Drosophila* and other fly larvae show regions where the chromosomes are puffed out and diffused. Analysis of the puffed regions shows that a large amount

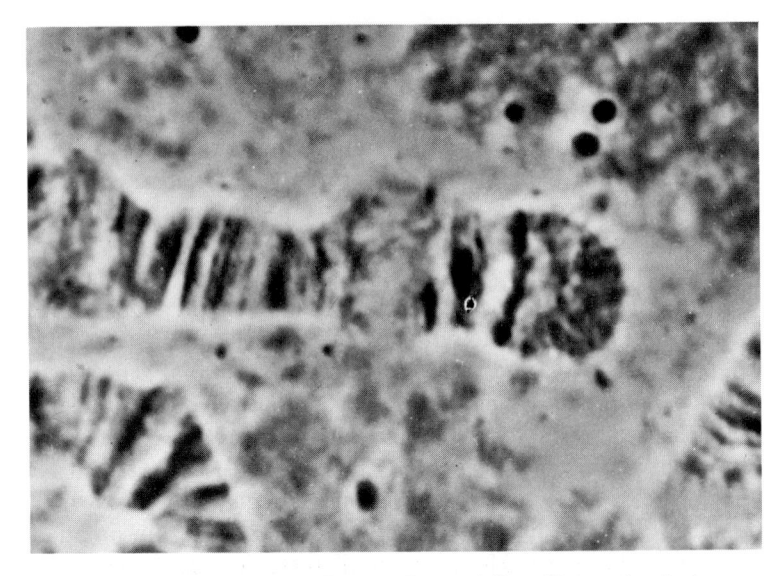

Fig. 15.7. Chromosome puffing in Drosophila salivary gland chromosomes. Evidence indicates that the puffing phenomenon is caused by an uncoiling of the DNA strands. This photograph shows a region in the center of the chromosome which is puffed out and indistinct as compared with the rest of the chromosome. Chemical analysis shows that there is a heavy output of RNA at the puffed regions.

of RNA is being synthesized here. It appears that the genes in the puffs have opened out and are producing RNA related to the activities of the cells in the salivary glands while the genes in the unpuffed regions are related to activities in other parts of the fly.

16: GENE MUTATIONS

A gene is an extremely stable unit—through countless thousands of replications a gene produces perfect copies of itself. Yet we know that in exceedingly rare cases mutations do occur, and the mutant gene replicates itself with the same exactness as its predecessor. Rare though they are, however, these mutations may have value. A New England farmer, Seth Wright, discovered a mutation among his sheep and from this established the ancon breed which is distinguished by its short legs. Navel oranges and seedless grapes are available today because of mutations which were discovered and propagated. More important than these practical applications, however, is the value of mutations in evolution. In a natural environment mutations provide new characteristics which lie at the base of natural selection and the improvement of a race. It will be our purpose in this chapter to learn more about mutations, how they occur, their frequency, the types of mutations, how they may be detected, and how they may be induced.

THE NATURE OF MUTATIONS

Sometimes the word mutation is used to refer to all types of genetic alterations, including chromosome aberrations; to avoid confusion the term is employed here in reference to those changes occurring within the functional gene units. These changes are sometimes called point mutations.

Early Studies of Mutations. In 1901, DeVries proposed the mutation theory as a result of his observations of the evening primrose. We now know that the variations he described were due to chromosome aberrations rather than point mutations, but his theories concerning mutations were essentially correct.

As with so many other important early genetic investigations, the first extensive mutation studies were carried out at Columbia University. There, T. H. Morgan found the white-eye mutation in a *Drosophila* stock which normally expressed red eyes. Spurred by this discovery, an intensive search was undertaken for more mutations. Millions of flies were bred and tediously examined, resulting in the discovery of over 500 mutations during the following 17 years. These included flies with purple eyes, rough eyes, small eyes, and no eyes; some had curved wings, bent wings, blistered wings, outstretched wings, miniature wings, and no wings; some had black bodies, yellow bodies, ebony bodies, and deformed bodies; there was even one that caused legs to grow out of the head instead of antennae. Stocks were established of these many mutations and much of the early genetic knowledge was obtained from studies using them.

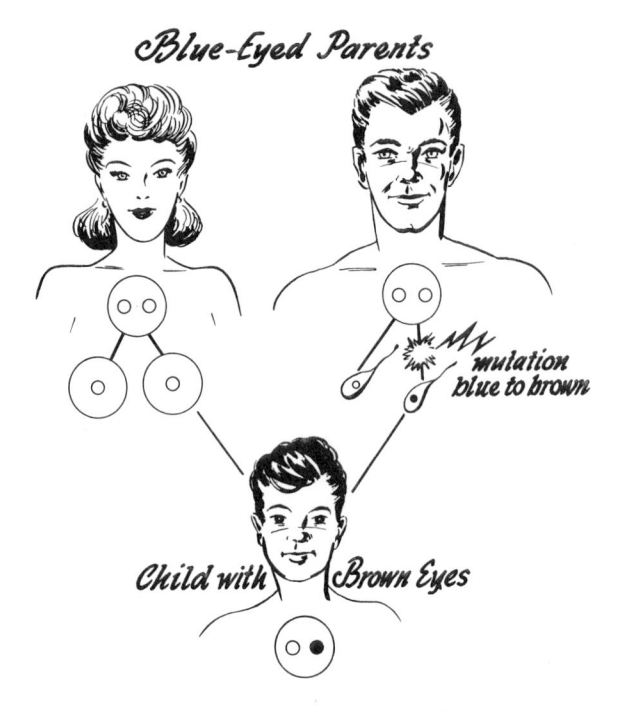

Fig. 16.1. A mutation in man. This diagram shows how a mutation can take place in a reproductive cell of a parent and result in a child with an inherited characteristic not in the germ plasm of either parent. (From Winchester, Heredity and Your Life, *Dover Press.)*

The Kinds of Mutations. From the preceding discussion one may receive the impression that most mutations bring about some striking, visible, phenotypic effect. As techniques for mutation detection have been refined, however, we have learned that the major physical changes which were found in the early mutation studies represent only a small part of the total number of mutations that take place. Many cause changes in physiological reactions which have no visible effect, but which may affect viability and fertility. The different kinds of mutations, together with their estimated percentage frequency, include the following.

1. DETRIMENTALS (80 per cent). These are mutations which have no visible effect, but cause some decrease in viability and or fertility of the organism expressing them. The expression of an individual detrimental gene may have only a slight effect, but the expression of a group of them in one organism can have a serious, even lethal, effect.

2. LETHALS (over 19 per cent). Most of the balance of the mutations have a physiological effect so extreme as to cause the death of a homozygous organism. These are the lethals.

3. VISIBLES (less than 1 per cent). This leaves only a fraction of 1 per cent of the mutations which cause some clearly detectable, visible phenotypic effect.

Reverse Mutations. In the early days of mutation studies, some theories stated that mutations represented destructive changes of the genes, perhaps the loss of a part, and that this accounted for the altered phenotypic expression. Then it was discovered, however, that mutation was not a one-way process. From large numbers of organisms expressing a gene which had arisen by recent mutation, there were very rare cases where one would undergo mutation that would restore the original phenotype. Such an occurrence is known as a reverse mutation. In *Drosophila* it was found that these generally do not occur so frequently as the mutations from the wild type to the mutant form, but the fact that they occur at all indicates that mutations are alterations, and not irreparable loses of genic material.

Harmful Nature of Most Mutations. From the above discussions it is evident that most mutations are harmful. Even visible mutations in most cases are harmful to some extent. The reason for this becomes clear when we think of the evolutionary process which has produced the living organisms on the earth today. A constant natural selection is eliminating from the populations the genes which have some harmful effect. In the fierce struggle for existence which goes on in most natural environments, the expression of a gene which lowers viability or fertility even to a small degree will be at a disadvantage and such a gene will tend to be replaced by genes producing a more vigorous constitution. This means that the genes found in a normal wild-type population are highly selected to retain the best of numerous mutations which may have appeared in the past.

It stands to reason that the alteration of anything that is already highly efficient will be more likely to do harm than good. We might compare the situation to an automobile engine which, although not perfect, runs in a satisfactory manner. Now if we raise the hood and make some change, without knowing what we are doing, it is much more likely that we will do harm rather than good. It is, of course, a very rare possibility, that we may just happen to hit upon a change which would make the motor run better. Likewise, there will be very rare alterations of genes which will make an organism more efficient in its particular environment. It is these rare mutations which make evolution possible. They will be selected for and eventually may become established as a part of the genotype of the majority of the population. The genes which are widespread in any species today had their origins as mutations at one time in the past. Hence, in this sense, all genes are mutant genes, although we commonly use the term to refer only to those genes which have undergone recent mutations.

Somatic Mutations. Genetic studies are concerned primarily with mutations which take place in the reproductive cells—these are the ones which will be transmitted to the offspring, but there is no reason why mutations cannot take place in any cell. Many of these probably do occur, but in a mature organism they are generally never expressed. A mutation affecting the color of the hair which took place in a cell of the intestine would have no chance of visible expression. There are a few cases, however, where there can be a phenotypic effect. Those which occur in very early embryonic stages can, through cell division, be propagated until a sufficiently large mass of tissue is formed to show the mutant phenotype. Mosaic effects can be produced when the mutant affects the body as a whole. *Drosophila* have been found with one half or one fourth of the body yellow and the balance the wild-type gray. This condition can be explained by a mutation to the sex-linked characteristic of yellow body in the early embryo. There are exceptional people who have eyes of different colors, perhaps one blue and one brown. This could be caused by a mutation in the early embryo. Of course, these conditions can also arise when there is chromosome deletion in heterozygous individuals.

Fig. 16.2. The ancon mutation in sheep. The ram on the left and the ewe on the right show the short legs and other characteristics resulting from a mutation of a gene in sheep. The ewe in the center has the normal dominant allele of this mutant gene. (Courtesy Life magazine.)

The variegated pattern which is sometimes seen on the leaves and flowers of plants can also arise by somatic mutation. In some plants there are very unstable genes which mutate frequently, making this condition common. In the larkspur, *Delphinium*, the gene for purple flower frequently mutates to white and yields a mixture of these colors on one plant.

Cancer represents an uncontrolled growth of tissue and there is much evidence that the disease can arise as a result of somatic mutation. Genes which influence the stability of the tissue and its correlation of growth with the other body tissue could mutate and cause a wild, uncontrolled growth. One piece of evidence favoring this concept is the fact that the same agents which induce mutation also can induce cancer. We know that heredity plays a part in the frequency of occurrence of cancer; perhaps this is due to the fact that some genes which influence the correlation of growth are more easily mutable than others.

METHODS OF MUTATION DETECTION

A good deal of the early experimental work in the artificial production of mutations failed because (1) the methods used were inadequate for detection of any mutations which might have been produced and (2) the number of offspring were insufficient to yield significant results. The methods vary for the different kinds of mutations.

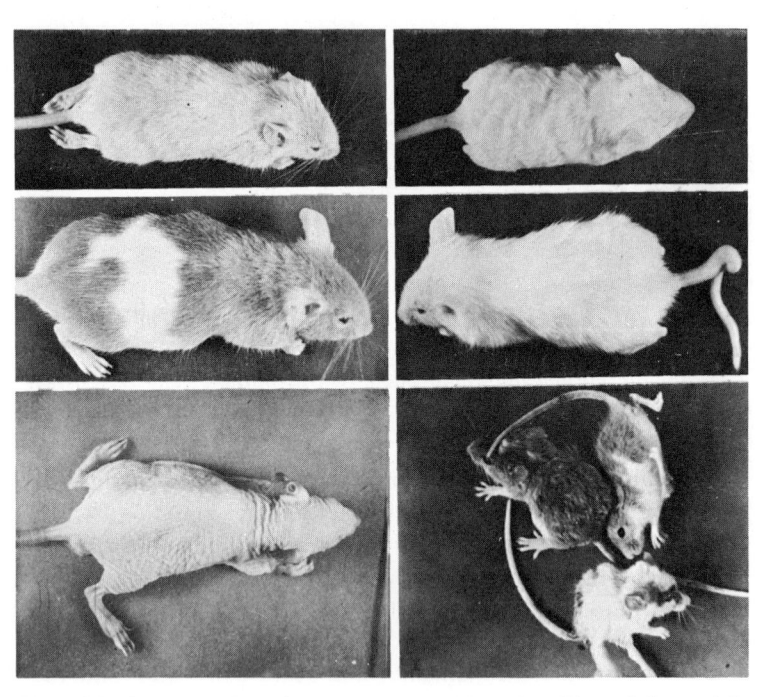

Fig. 16.3. Some results of gene mutations in mice. From left to right and top to bottom these are: short ear, waved fur, belted body, kinky tail, hairless body, and jittery nervous system. (Courtesy George Snell, Jackson Memorial Laboratory, Bar Harbor, Maine.)

Detection of Visible Mutations. If visible mutations have a dominant or intermediate phenotypic effect, they are easy to detect because they will be expressed in the first individuals containing them in their somatic cells. Recessives offer a greater problem because they can be carried for many generations without being expressed. The simplest means of detecting these is close inbreeding. If an individual receives a recessive mutation from a parent, he will transmit it to about one half of his offspring. Thereafter, either an *inter se* cross of the offspring or a back cross will effectuate the phenotypic expression of the gene in the third generation of offspring (since the occurrence of the mutation), providing sufficient numbers of offspring are obtained.

Sex-linked recessive mutations are somewhat easier to detect because they can appear in the first or second generation without the necessity of special inbreeding. In organisms employing the XY method of reproduction, the male offspring will show mutations which occurred in the female parent. The second generation males may express any mutation that took place in the male parent.

Special techniques have been devised to render the method of mutation detection more effective. One of these techniques allows detection of mutations at specific loci in the first generation offspring. Parents are selected that are homozygous for a number of recessive genes; let us assume for purposes of illustration that we use female mice with ten such genes. We breed these to males which may or may not have been treated in some way in an effort to increase the number of mutations. If a mutation has taken place in the reproductive cells of the males which is the same as any one of the recessive genes carried by the female, then an offspring may be produced which will be homozygous; thus this particular gene will be expressed. The other nine genes will be heterozygous and will not be expressed. The primary difficulty with this method is that we are limited in the number of recessive genes which can be incorporated into one stock. Therefore, we overlook many mutations which take place at other loci and thus, we must obtain large numbers of offspring in order to acquire significant data. In the laboratories at Oak Ridge, Tennessee, where many important studies of mutation in mammals (especially mice) have been made, it is common to raise as many as 500,000 mice in one such investigation. In *Drosophila* it is much simpler to deal with such great numbers, but the work remains tedious.

A chromosome aberration, known as the attached X, has made possible the detection of sex-linked visibles in *Drosophila*. In flies with this aberration the two X-chromosomes of the female are attached at one end and they do not separate in meiosis. Thus,

220 GENE MUTATIONS

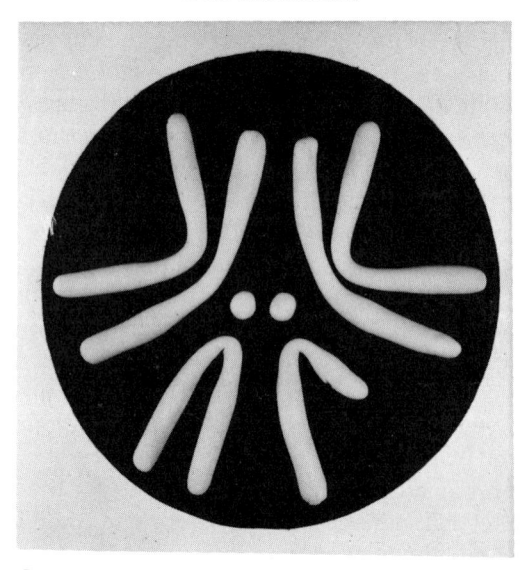

Fig. 16.4. Attached X-chromosomes. This is a model of the chromosome complement of the diploid cells of a female with the X-chromosomes fused together at one end. This results in some eggs with 2 X-chromosomes and some with none. The Y-chromosome is present, but plays no part in the determination of sex of the offspring.

there is always non-disjunction of the X-chromosomes—one half of the eggs are XX while the other half have no X-chromosome. It will be noted, from the diagram of this cross in Fig. 16.4., that all the male offspring receive their X-chromosome from the male parent rather than the female parent as in the normal oögenesis. Should any mutations occur in the X-chromosome of the male parent they will be expressed in the male offspring in the first generation. This stock of flies is especially valuable when one wishes to test methods of increasing mutation frequency.

Detection of Lethal Mutations. Mutations which have a lethal effect when homozygous can be detected rather easily if there is some phenotypic expression when they are heterozygous. The unusual 1:2 ratio obtained when two such heterozygous individuals are crossed is characteristic of this type of lethal. (See Chapter 4 for details of this type of cross.)

Since most lethals have no easily detectable phenotypic effect in the heterozygote, we must devise other methods of detection. We can use the inbreeding technique described for visible mutations, but we must note carefully the absence of certain individuals rather than the presence of unusual forms. A recessive lethal mutation which appears in a germ cell will be heterozygous in the individual formed from that cell. Some of its offspring will also be heterozygous. In the third generation there will be some matings of two heterozygous individuals. About one fourth of their offspring will die. If the organism employed is a plant, about one fourth of the seed will not germinate, or if it does germinate it will produce a plant which cannot survive. In oviparous animals

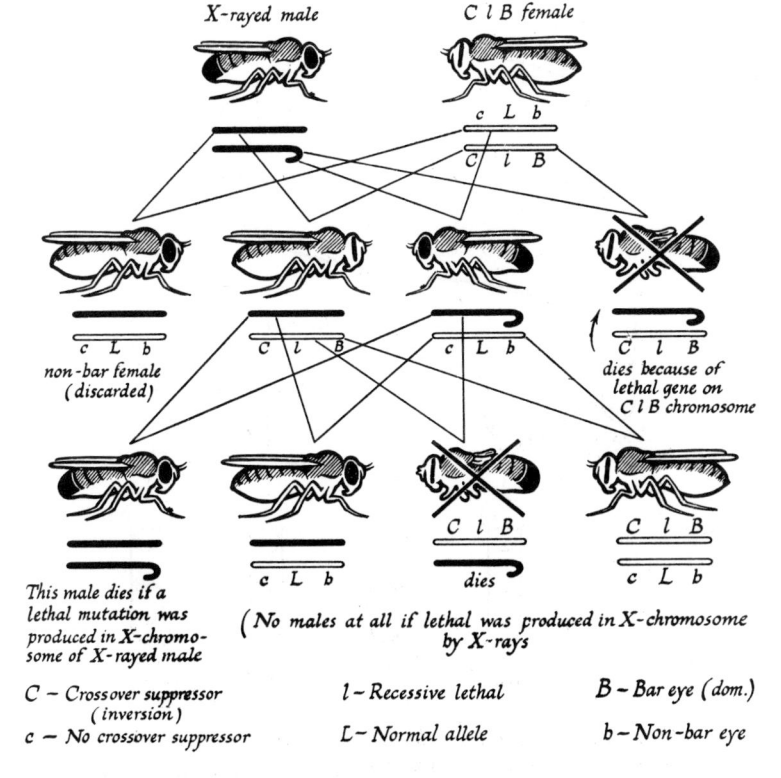

Fig. 16.5. The ClB method of detecting lethal mutations in Drosophila. *In practice, an unrelated male is usually selected to mate with the F_1 female.* (From Winchester, Genetics, Houghton Mifflin.)

one fourth of the eggs will not hatch, or will hatch into offspring which will die shortly. In mammals about one fourth of the offspring will die before, or shortly after, birth. If the organism survives until a fairly late stage of embryology, it can be recognized rather easily. The lethals causing earlier deaths must be recognized by means of a statistical analysis of the offspring.

Sex-linked lethals may be recognized by the unusual sex ratio. In the XY organisms a heterozygous female will have offspring in a 2:1 ratio in favor of females because one half of the males will die. The heterozygous female escapes detection, however, until several generations after a mutation has occurred. A much more

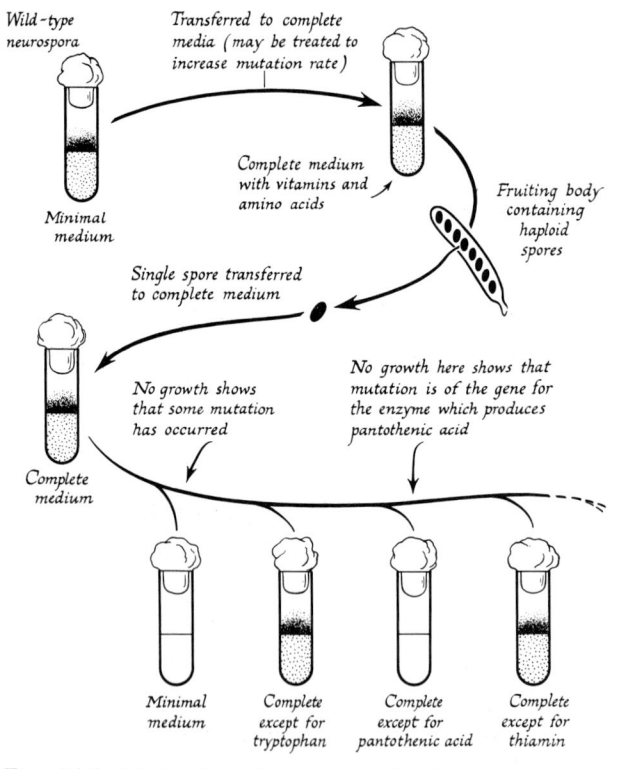

Fig. 16.6. *Method of detection of biochemical mutations in* Neurospora. *This mutation results in an inability to synthesize pantothenic acid from minimal food substances.* (From Winchester, Genetics, Houghton Mifflin.)

precise method of detecting sex-linked lethals in *Drosophila* was worked out by H. J. Muller of the University of Texas while doing his pioneer work on the induction of mutations. His method is known as the *ClB* method. *ClB* females have one normal X-chromosome and one X-chromosome in which the central portion has been inverted. This inversion prevents crossing over, and the C in the name given to such chromosomes refers to this crossing over suppressor. This unusual chromosome also carries a lethal, represented by the *l*. Finally, there is a dominant gene for bar eye, *B*, which acts as a marker to identify the chromosome. Females heterozygous for this chromosome are crossed to wild-type males which may or may not have been treated by some agent which might increase the mutation rate. The results of this particular type of cross are shown in Fig. 16.5. Muller's method demonstrates that it is a simple matter to detect a lethal. There will be no male offspring in the second generation cultures if a lethal occurred in the male being tested. A large number of cultures can be prepared and easily checked as a quantitative measure of the frequency of sex-linked lethals.

Mutation Detection in Haploid Organisms. Since all genes act as dominant genes in haploid organisms, the latter are especially favorable for mutation studies. We have already mentioned some of the techniques used in the mold *Neurospora* and in bacteria. By use of the proper culture media, biochemical mutants as rare as one in each ten billion organisms can easily be detected. In some of the insects there is the added advantage of having haploid males; hence all genes act as sex-linked genes and all mutations can be detected as easily as sex-linked mutations. The small parasitic wasp, *Habrobracon*, has been employed for many such genetic studies.

MUTATION FREQUENCY

The rate at which mutations occur varies according to species and other factors. Since they are comparatively rare, it is difficult to obtain accurate estimates of frequency; however, with certain forms considerable information has been obtained.

Drosophila. Dr. H. J. Muller, mentioned above for his pioneer studies in *Drosophila*, estimates that any single gene has only about one chance in a million of undergoing mutation during its life span from one replication to another. This indicates the stability of the gene and the accuracy of the replication process, yet when we consider the number of genes in the *Drosophila* cells and the number of replications that can exist, we find that there can occur about one mutation for every 20 reproductive cells produced. This figure is much higher than is generally conceived because it includes all of the minor changes which ordinarily are not detected. Only a fraction of 1 per cent of these would be visible.

Fig. 16.7. Chondrodystrophic dwarfism, a dominant trait which appears in about one child out of every 11,500 born to normal parents as a result of mutation. (Courtesy C. Nash Herndon.)

In Man. The mutation rate for *Drosophila* is at least ten times that for man per generation. Considering the great difference in the length of generations, the rate is about 500 times greater for any given period of time. Because of man's long generation cycle (approximately 30 years), he could not tolerate a mutation rate equal to that of *Drosophila*. The accumulation of harmful mutations over this long period would cause the extinction of the species. Haldane estimates that any single human gene has an average life of 2,500,000 years before it is changed by mutation. Such an estimate has been obtained by a study of the mutations which appear among children born in hospitals. For example, detailed records from the hospitals in Copenhagen reveal that out of 128,763 births there were 14 children who were chondrodystrophic dwarfs. This is a condition in which the head and trunk are of normal size, but the legs and arms are very short. It is caused by the presence of a dominant gene. Three of these children had parents with the condition but the other 11 descended from normal parents. The latter cases must represent mutations. This is a rate of one in about 11,700

Mutations in Man

Mutation	Effect on Body	Frequency: Once in Every	Mutation Rate in Percentage
DOMINANT			
Pelger anomaly	White blood cells abnormal—disease resistance reduced	12,500 gametes	0.0080
Chondrodystrophic dwarfism	Shortened and deformed legs and arms	14,300 gametes	0.0070
Retinoblastoma	Tumors located on eye retina	43,500 gametes	0.0023
Aniridia	Absence of iris of eye	200,000 gametes	0.0005
Epiloia	Red lesions on face; tumors in brain, kidneys, and heart	83,333 gametes	0.0012
RECESSIVE (autosomal)			
Albinism	Melanin undeveloped in hair, skin and iris	35,700 gametes	0.0028
Amaurotic idiocy (infantile)	Deterioration of mental faculties during first months	90,909 gametes	0.0011
Icthyosis congenita	Rough, scaly skin at birth, may cause infant death	90,909 gametes	0.0028
Total color blindness	Difficulty distinguishing colors	35,700 gametes	0.0028
RECESSIVE (sex-linked)			
Hemophilia	Blood clots very slowly	31,250 gametes	0.0032

births or one in every 23,400 genes at this locus in the parents. If we take the average age of parents as thirty years at the time of conception, we find that this particular gene mutates about once every 690,000 years; this is somewhat more frequent than the estimate given by Haldane, but the gene might be of the type which mutates more readily than some of the others. We know that genes do vary in their rate of mutation, and a study of other human genes shows that this is true in man. The table on page 219 gives some of the figures which have been compiled from Haldane, Neel, Shull, and Falls.

Of course, it must be realized that these figures are obtained from a comparatively small number of mutations and, therefore, the percentages would be subject to rather large standard deviations, although the figures do give an indication of the mutation rate in man.

Variations of the Mutation Rate in Maize. The commercial corn, maize, has been studied extensively by L. J. Stadler, of the University of Missouri. He compiled a list of the different mutants and found that there was considerable variation at the different loci. In some cases this was as great as a hundredfold. His results are given below.

MUTATION RATES OF GENES IN MAIZE

Mutation	Frequency: Once in Every	Mutation Rate in Percentage
R—a color factor for leaves and aleurone	2,008 gametes	0.0492
in—intensifier for color	9,500 gametes	0.0106
pr—red aleurone	92,350 gametes	0.0011
su—sugary endosperm	419,000 gametes	0.00024
Y—yellow endosperm	434,000 gametes	0.00022
sh—shrunken endosperm	823,000 gametes	0.00012
wx—waxy endosperm	none found in 1,503,744 gametes	

Variations Due to Temperature. In cold-blooded animals, temperature can be an important factor in the mutation rate. This is to be expected—heat accelerates chemical reactions and, basically, mutations are chemical alterations. Muller and several other workers found that the number of mutations in *Drosophila* would be accelerated by as much as fivefold if the flies were kept for their entire lives at a temperature 10 degrees Centigrade higher than control flies. For warm-blooded animals this would not generally be true since they maintain a constant body temperature, although an Atomic Energy Commission official did receive headline newspaper reports when he facetiously suggested that hot baths and tight underwear for men might cause more mutations than radioactive fallout.

Variations Due to Sex. Studies in *Drosophila* indicate that sex is a factor in the rate of mutation. Using the sex-linked mutations because of their convenience of identification, it was found that they occurred more frequently in males than females. From the studies of hemophilia there was an indication that the same is true of human beings.

THE ARTIFICIAL INDUCTION OF MUTATIONS

Since the discovery of mutations, their potential value has been recognized. To the commercial plant and animal breeder they have offered an opportunity to find new characteristics which might be selected to improve existing varieties. To the theoretical geneticists they have extended an opportunity to learn more about the nature of the gene and its functions. Natural mutations are so rare in occurrence, however, that it was soon after their discovery that geneticists were trying to find ways to speed up the process.

Mutation Induction by X rays. The first break-through in the search for a means of mutation induction came in 1927 when Dr. H. J. Muller (who had been trained by the pioneers at Columbia University and was then working at the University of Texas), found that the mutation rate could be greatly increased through the exposure of *Drosophila* to X rays. Realizing that lethals were more numerous than visibles, he used the *ClB* method of detecting lethal mutations. Through heavy doses of radiation applied to the male parents, he obtained a hundredfold increase in the number of mutations.

Similar experiments on barley were being carried out by L. J. Stadler in Missouri, but, unfortunately, the longer generation cycle of barley required two years for detection of mutations and thus

Dr. Stadler's results appeared a year after Muller's findings on *Drosophila*.

Since these discoveries the X-ray machine has become a tool of standard equipment in genetic research laboratories because of its great value in increasing the rate of mutations and, as was later discovered, chromosomal aberrations.

Were these induced mutations any different from those which occurred naturally? This was an important question for it was possible that the changes brought about were only destructive changes. Extensive studies, however, have shown that they are not different. The visibles, lethals, and detrimentals, so far as can be determined, take place in the same proportion—there are just more of them following the radiation. It appears to be simply a way of speeding up a natural process.

Radioactive Elements. Shortly after it was discovered that X rays could induce mutations, investigations began on the radiation given off by radium. It was found that the "hard" gamma rays would induce mutations in the same proportion as X rays, according to dose. In recent years, the development of the knowledge of atomic fission and of the great host of radioactive isotopes which can be produced has increased the importance of radiation as a source of mutation. A further discussion of this significant topic can be found in Chapter 17.

Chemical Agents. Since mutation is essentially a chemical alteration, various chemicals have been tested as possible mutagenic agents. However, this presented the problem of undue exposure of the reproductive cells—an experimental animal usually would be killed if the entire body were exposed to the chemical in sufficient concentrations to have any possible effect. This complication was avoided by removal of the gonads and immersing them in the chemical and then transplanting them to another animal. In addition, eggs have been exposed after laying.

Of the many chemicals tried, positive results have been obtained with mustard gas, nitrogen mustard, ethyl urethane, phenol, and formaldehyde. A number of these chemicals have been found to be carcinogenic (cancer producing).

Ultraviolet Rays. Ultraviolet rays have been found to be mutagenic in certain cases. These rays have longer wave lengths than the X rays; therefore they are not as penetrating and will not affect dense tissue. The bacteria and molds, however, are sufficiently small to absorb the radiation, and significant results have been obtained by treating these organisms. In addition, important results have been obtained with selected portions of larger organisms. The pollen grains of maize and the polar cap region of the egg of *Drosophila* are examples of these.

One of the most significant findings stemming from the ultraviolet ray studies is that there is a difference in the mutagenic effects of different wave lengths of the ultraviolet spectrum. Some lengths are more effective than others, and the effectiveness is in direct proportion to the absorption of these rays by DNA. Those which are most readily absorbed cause the greatest number of mutations. This observation can be cited as evidence in favor of DNA as the true genetic material.

Of the different mutagenic agents which have been discovered, only radiation of high energy and short-wave length seems to be of any possible significance as far as mutations in man are concerned.

CAUSE OF NATURAL MUTATIONS

When it was discovered that radiation could induce mutations, the question naturally arose concerning the cause of natural mutations. Could it be that the natural radiation to which we are all exposed is the cause of the mutations that occur regardless of any other treatment? Cosmic rays from outer space that come down upon us constantly fall in the category of high-energy radiations. Could they be a source of natural mutations? In addition, a small part of many natural elements are radioactive—the air we breathe contains a small number of radioactive carbon atoms (C^{14}). The bricks or stones in a home or building contain radioactive elements; so does the ground upon which we walk. There is evidence that this natural background radiation is the cause of some of the natural mutations. For example, on Pike's peak at an altitude of 14,000 feet, the intensity of the cosmic rays is 15 times greater than at sea level. A significant increase in mutation rate has been found in *Drosophila* raised at this altitude as compared to those raised at sea level.

Calculations of the amount of radiation from natural sources to which *Drosophila* would be exposed, however, reveal that it is not sufficient to account for the number of natural mutations which occur. We can only speculate on the cause of the others. Perhaps a substitution of a purine or pyrimidine in the replication of the DNA, perhaps energy fluctuations which could cause some rearrangement within the chain, or perhaps some other cause about which we have no information at present is the key factor in mutation causation.

THE HARDY-WEINBERG PRINCIPLE OF DETERMINING FREQUENCIES OF MUTANT GENES

It is often desirable to know the frequency of certain recessive mutant genes in a population and the number of heterozygous carriers of the gene. For instance, if a person knows that a certain harmful recessive gene is in his family, it may be important to him to know his chances of marrying a person who carries the same gene. This information can be obtained by application of the *Hardy-Weinberg principle.* To use this principle one must first know the approximate frequency of the persons in the population who are homozygous for the recessive gene. This information can be obtained by surveying a representative sample of the population and determining the frequency of those who express the recessive trait. The square root of this figure will be the frequency of the recessive gene in the population, and from this we can figure the frequency of homozygous and heterozygous dominant persons in the population.

To illustrate, let us consider a common human trait. We know that a certain dominant gene gives a person the ability to roll the tongue (Fig. 4.2). Let us assume that a survey of a typical sample of the people in your city shows that $1/16$ of them cannot roll their tongues. This means that they are homozygous for the recessive gene. To apply the Hardy-Weinberg principle we must refer back to some of the principles of probability discussed in Chapter 6. Since each person has two genes related to tongue rolling, we can use $(a+b)^2$ to determine the frequencies of each of the two alleles. Let $a=$ the frequency of the dominant allele (R), and $b=$ the frequency of the recessive allele (r). Expanding the binomial we get:

$$(a+b)^2 = a^2 + 2ab + b^2$$

We know the value of b^2; this is the fraction of the population which cannot roll their tongues (they are homozygous for the recessive gene.) So we can determine the frequency of the gene r by taking the square root of b^2.

$$b = \sqrt{b^2} = \sqrt{1/16} = 1/4$$

The rest of the alleles must be R, so we obtain the frequency of this dominant allele by subtraction from one (the total number of genes).

$$a = 1 - b = 1 - 1/4 = 3/4$$

We can now determine the number of heterozygous carriers of the gene:

$$2\,ab = 2 \times 3/4 \times 1/4 = 6/16$$

The homozygous dominants would be:

$$a^2 = (3/4)^2 = 9/16$$

Thus, starting only with the knowledge of the frequency of the homozygous recessives we obtain values for the heterozygous and homozygous dominants as well as the frequencies of both alleles in the population.

In many cases, especially those wherein the recessive genes are of very low frequency, it is more convenient to use percentages or decimals rather than fractions. This same problem solved by the decimal method would be:

$$rr \text{ (or } b^2) = .0625 \text{ (or } 1/16)$$
$$r = \sqrt{.0625} = .25$$
$$R = 1.0 - .25 = .75$$
$$Rr = 2 \times .75 \times .25 = .3750$$
$$RR = (.75)^2 = .5625$$

In this application of the Hardy-Weinberg principle we have assumed that there is no selective advantage of one gene over another. If we work out the mathematical proportion of persons in Africa who are heterozygous for the gene for sickle-cell anemia, based upon the number of persons who have the anemia, our results will be too low because many of the homozygous persons die before the tabulation is made. Hence, the number of homozygous persons with the anemia is not a true indication of the frequency of the gene in the population. The divergence is accentuated by the fact that the heterozygous persons have an advantage in malarial resistance over those homozygous for normal hemoglobin. When such facts are kept in mind, the Hardy-Weinberg principle can be a valuable means of estimating gene frequency.

17: RADIATION HAZARDS IN AN ATOMIC AGE

Today man finds himself in an age in which there are ever-increasing sources of high-energy radiation. In view of the discovery that such radiation can produce mutations and chromosome aberrations, there is growing concern on the part of many that damage to future generations may occur as a result of increased exposure to such radiation. There are diversities of opinion regarding the degree of danger involved, although geneticists as a whole are generally agreed that it is a matter of grave concern. In this chapter we will review the sources of radiation to which man may be exposed and we will try to evaluate the possible consequences of exposures in the light of the most authoritative opinions available.

SOURCES OF HIGH-ENERGY RADIATION

In 1895, the German physicist Roentgen discovered the penetrating rays known as X rays. Since then many other sources of high-energy radiation have been discovered and made available for man's scientific and practical utilizations. We shall review some of these briefly.

X rays. X rays are formed when electrons from a high voltage electric current jump across a gap within a vacuum tube. As the electrons strike a platinum target, energy is given off in the form of short-wave, high-energy radiation, i.e., X rays. Because of their penetrating quality, these rays have proved to be of great value to mankind in medical diagnosis—one can literally "see" inside the body. While the human eye is unable to make this observation directly, photographic film is sensitive to these rays and one can make photographic images of the internal body structure. Not all of the rays are absorbed—the denser body structures absorb proportionately more than the less dense organs; this factor produces a shadow image on the film. In addition, a screen containing a chemical which shows fluorescence when

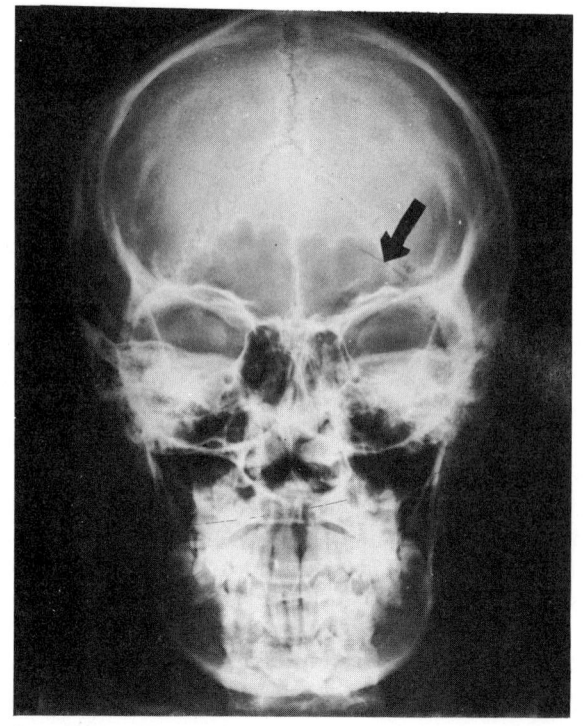

Fig. 17.1. *The medical value of X rays is illustrated by this picture of a human head showing the location of a fracture of the skull, indicated by the arrow.*

struck by the X rays may be placed in the path of the rays, thus making the image visible to the human eye.

Radioactive Chemical Elements. The majority of atoms that comprise the substance of the earth have a stable balance of their intra-atomic structure; there are some, however, which are unstable—they tend to undergo change, thereby achieving stability. These unstable atoms were discovered within a few months after the discovery of X rays. Henri Becquerel, a French scientist, noted that uranium ore gave off penetrating radiation. This occurrence was proved when he placed some of the ore near a photographic film and upon development the film showed a darkening of those regions which had been placed near the uranium. This discovery prompted Pierre Curie and his wife, Marie, to attempt to isolate the radioactive substance from the ore, thus leading to the discovery and isolation of radium. Since that time a very large number of radioactive elements have been discovered. To understand these we must learn something about atomic structure.

Atoms are the tiny, ultra-microscopic units of which elements are composed. An atom consists of a rather dense **nucleus** around which revolve small, negatively-charged bodies known as **electrons.** Within the nucleus there are positively-charged particles known as *protons*, which are very large in comparison to electrons—they weigh about 1,800 times as much. The number of protons and electrons are equal, thereby creating a condition of neutrality. For example, hydrogen has one of each, oxygen has eight of each, carbon has six of each, and uranium has ninety-two of each. In addition, in the nucleus of all except the lightest of elements (hydrogen) there are particles without charge known as **neutrons.** They have a weight about equal to that of protons and, therefore, add weight to the atom without altering its charge. Moreover, they play a part in the stability of the atom. The atom contains a tremendous amount of bound energy; when the number of neutrons is altered there may arise such an unstable state that the

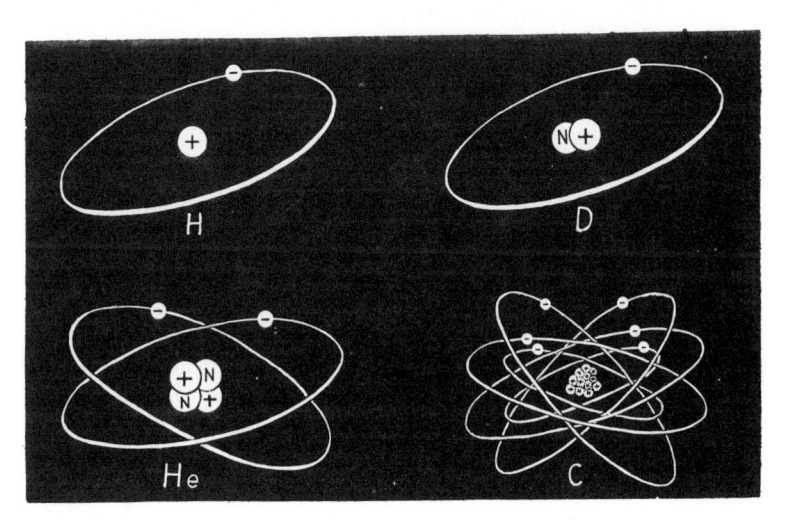

Fig. 17.2. The structure of some atoms according to the generally accepted theory. Electrons are indicated by a minus sign, protons by a plus sign, and neutrons by N. Atoms shown are hydrogen, deuterium, helium, and carbon. (From Winchester, Biology and Its Relation to Mankind, Van Nostrand.)

atom tends to split or to give off particles or energy, thereby achieving stability. Such unstable atoms are known as *radioactive isotopes*. For example, the normal carbon atom has six protons and six neutrons in its nucleus and six electrons in orbit around this nucleus. This atom is called carbon twelve, C^{12}, but there is one form of carbon which has eight rather than six neutrons. This gives the atom a weight of fourteen rather than twelve and hence it is known as C^{14}. The extra neutrons create a state of instability and the heavy carbon atom tends to give off radiation.

Radiation from radioactive isotopes falls into two categories. *Particulate radiation* consists of actual subatomic particles which come off from the atom with rather high energy. *Electromagnetic radiation* is in the form of short-wave, high-energy radiation similar to X rays. The characteristics of the different types of radiation from isotopes follow.

Alpha Particles. These are by far the heaviest of the particles which are given off in the aforementioned changes. These particles consist of a combination of two protons and two neutrons. Because they are positively charged, they are slowed down readily by the negative charges in matter and therefore do not penetrate very deeply into living tissues.

Beta Particles. Electrons emitted from atoms at high energy are known as beta particles. They vary in their penetrating power because they express considerable variation in the energy with which they are emitted. Because they are negatively charged, they are slowed by positive charges and lose energy rapidly. Hence, they are not very penetrating.

Neutrons. Free neutrons may be emitted in radioactive isotope decay. These neutrons may be extremely penetrating because, lacking any electrical charge, they are not slowed by passing near charged particles, and thus tend to move in a straight line until they have a collision. Since we know that even the heaviest of atoms consists largely of empty space, the neutrons may travel a considerable distance before such a collision takes place.

Gamma Rays. The gamma rays are of short-wave length—high-energy rays of the same nature as those given off from a high-voltage X-ray tube. They represent energy released from the atom in the form of electromagnetic radiation. When a uranium atom undergoes fission, its daughter products have a mass slightly less than that of the parent atom. This lost mass has been converted into energy in accordance with Einstein's famous equation $e = mc^2$ (energy in ergs is equal to the mass times the velocity of light in centimeters squared). Some of this energy is converted into light and heat, but much of it goes into the production of gamma rays which are very penetrating.

Cosmic Rays. The complex, high-energy radiation known as cosmic rays showers down upon us constantly. These rays have been detected in salt mines at a depth of about 2,000 feet. This fact testifies to their highly penetrating properties. Primary cosmic radiation comes to us from outer space in the form of stripped nuclei of such elements as carbon, nitrogen, and oxygen. Few of these reach the earth, for our atmosphere acts as a shield with a density equivalent to three feet of lead. Some of the rays, however, collide with nuclei in the atoms of the atmosphere and give rise to a shower of both particulate and electromagnetic radiation known as secondary cosmic rays. This material forms an important part of the inescapable radiation to which all life on earth is exposed—the background radiation.

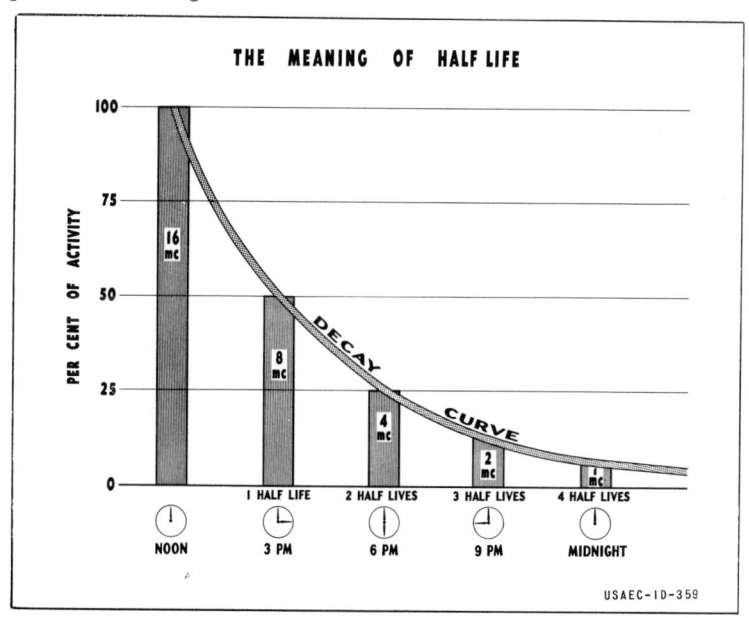

Fig. 17.3. Rate of decay of radioactive elements in terms of half life. This example is based on a half life of three hours. (Courtesy United States Atomic Energy Commission.)

The Life of Radioactive Isotopes. Since radioactive isotopes achieve stability as they give off radiation, it would be expected that a mass of radioactive material would, in time, lose its radioactivity. This is true, but the loss of radioactivity does not follow a linear course. Radioactive decay is commonly expressed in terms of half life. For example, the isotope of iodine, I^{131}, has a half life of about 8 days. This means that any given volume of this isotope will lose one half of its radioactivity in 8 days. In the next 8 days it will lose one half of that which is remaining and so on. Thus, in 16 days it will be one fourth as active as it was at the beginning, in 24 days it will be one eighth as active, and in 32 days it will be only one sixteenth as active.

The rate of radioactive decay varies greatly for different isotopes, some having a half life of as little as a fraction of a second while others run into many thousands of years. For each radioactive element the rate of emission of radiation takes place at a specific half-life rate. Just why the atoms in a given mass of radioactive iodine do not all give off radiation at once and achieve stability in one "fell swoop" no one knows; we only know that a certain proportion of those in the mass achieve stability in any particular second. To show the wide variation in half lives and in the type of radiation given off by different isotopes, several of the most common are listed below.

SOME RADIOACTIVE ISOTOPES

Isotope	Radiation Emitted	Half life
Nitrogen 16	beta and gamma	7.4 seconds
Sulfur 37	beta and gamma	5 minutes
Sodium 24	beta and gamma	15 hours
Gold 108	beta and gamma	2.7 days
Iodine 131	beta and gamma	8 days
Iron 59	beta and gamma	45 days
Cobalt 60	beta and gamma	5.2 years
Strontium 90	beta	28 years
Radium 226	alpha and gamma	1,620 years
Carbon 14	beta	5,600 years
Chlorine 36	beta	310,005 years
Uranium 235	alpha, beta, gamma and neutrons	710 million years

MEASUREMENT OF RADIATION

No type of high-energy radiation previously described can be detected by any of man's senses. A piece of radioactive cobalt (Co^{60}) the size of a pea could be affixed to the bottom of the seat of a chair and a person sitting in the chair could receive a lethal dose within a few minutes without ever feeling, seeing, hearing, tasting, or smelling anything that would indicate the presence of the isotope. Hence, scientists must employ instruments for measuring purposes.

Fig. 17.4. Use of the Geiger tube and a pulse counter in measurement of radiation.

Measuring Instruments. The best-known and most widely used instrument for measuring isotopes is the **Geiger-Müller** tube. Radiation entering the tube causes ionization of the gas within; this condition creates a pulse which can be detected with a suitable pulse counter. This method is accurate for measuring gamma radiation and high-energy beta radiation, but of little value in determining alpha radiation because the alpha particles do not penetrate its window. The *proportional counter* and the *ionization chamber* are other instruments based on this property of ionization. The *scintillation chamber* is based on the principle that radiation causes flashes of light when it strikes certain fluorescent materials. It can be used to measure alpha particles as well as beta and gamma rays.

Standards of Measure. The most widely used standard of measure of radiation is the roentgen (*r*). This method was adopted by the Radiological Congress of 1937 as a standard for expressing quantities of electromagnetic radiation. Roentgen is expressed as the amount of radiation that will produce one electrostatic unit of charge in one ml. of air at standard temperature and pressure. This innovation was an important standardization; previously the quantity of radiation had been stated frequently as the number of seconds of exposure at a certain distance from an X-ray tube of a certain voltage—such a measure introduced too many sources of error.

As the development of the study of particulate radiation progressed, however, it became necessary to use other terms since the amount of ionization in gas was not in proportion to its effect on tissue. The *rep* (roentgen equivalent physical) was introduced in an effort to express particulate radiation in terms of its biological effect, in terms comparable to those used for the electromagnetic radiation. This standard of measurement is being replaced by the *rad* (radiation absorbed dose) which is based upon the energy absorbed per gram, either in living or nonliving material. Finally, the *rem* (roentgen equivalent man) has been used in studies of

UNIT MEASUREMENTS OF IONIZING RADIATION

Unit	Energy Absorption (ergs per gram)	Measured in	Radiation Measured	General Use
Roentgen	83	Air	X & gamma	Monitoring
Rep	93	Soft tissue	Particulate	Biological research
Rad	100	Any material	All types	Biol. and physical research
Rem	—	—	All types	Records of human exposures

radiation effects on man in an effort to compile comparable figures for the great variety of radiation to which man can be exposed in this atomic age. It is the quantity of radiation which produces the same biological effect on man as that which results from the absorption of one roentgen of electromagnetic radiation. The above table summarizes these different standards of measure.

BIOLOGICAL EFFECTS OF HIGH-ENERGY RADIATION

The effects of radiation on the human body may be classified under two headings: the biological, which are the effects on the individual exposed; and the genetic, which are the effects that will be transmitted to future generations because of changes occurring in the reproduction cells. Although this book is primarily concerned with the genetic effects, we shall survey the effects on the individual because (1) these aid in our understanding of the genetic effects and (2) it is of vital importance that these effects are understood by everyone.

Ionization Induced by Radiation. High-energy radiation, both particulate and electromagnetic, is often spoken of as ionizing radiation because of the ions it produces in the matter through which it passes. An ion is an atom with either a positive or a negative charge. Radiation can produce ionization whenever a fast-moving, charged particle passing through matter pulls an electron out of the orbit of an atom. The atom then becomes an ion with a positive charge, due to the fact that there will be one extra proton in the nucleus that is not balanced by an electron in the orbit. The free electron is usually captured in the orbit of another atom; this process converts it into a negative ion. Neutrons, being without charge, do not cause the expulsion of orbital electrons in this manner, but they can cause ionization by a less direct means: a neutron may strike the nucleus of an atom. This occurrence causes excitation and the emission of a charged particle; the emission, in turn, will produce ionization by affecting orbital electrons. Electromagnetic radiation also causes ionization in a secondary manner. When the energy of the radiation is absorbed an atom is said to be in an excited (unstable) state. The excess energy which has been absorbed is dissipated by throwing off one of the orbital electrons. This electron produces additional ionization since it is a charged particle moving through matter. One roentgen of radiation produces about two billion such ion pairs in air.

Effect of Ionization in Living Tissue. It is believed that the damage done by radiation to living tissue is largely brought about by ionizations. An active cell is composed mostly of water thus, the radiation is most likely to affect the water by producing ions and free radicals which combine with oxygen, thereby producing

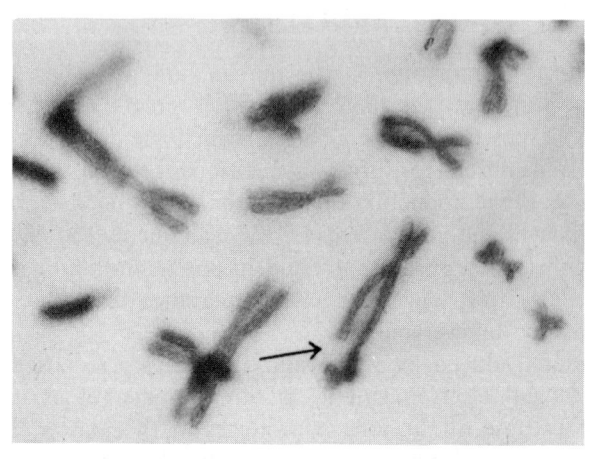

Fig. 17.5. Aberration in human chromosomes induced by radiation. Human cells growing in tissue culture were exposed to 50 r of X rays. An average of 20 per cent of the cells showed breaks such as those seen here. (Courtesy Michael Bender, Johns Hopkins.)

chemicals which are quite reactive. These chemicals may react with enzymes and other vital protein constituents in the cell of living tissue; they may also react with the chromosomes, thus causing chromosome aberrations and mutations in the genes. Any of these changes can bring about (1) the death of a cell or (2) an alteration of its pattern of growth and reproduction in such a way as to make it abnormal. Much of the long delayed radiation damage manifests itself as cancer of the tissues, caused initially by these changes. The cancers have been known to appear as long as 30 years or more after administration of the radiation. It is ironic to note that the valuable tool which is used to destroy cancer can also, when used improperly, induce the disease.

Differential Effect of Radiation. Not all cells are equally sensitive to damage by radiation. In general, those cells which are actively dividing are much more sensitive than those that are not dividing. In addition, cells in an active stage of metabolism are more sensitive than those which are relatively inactive. The blood-forming cells of the bone marrow, the actively dividing cells of the skin and intestinal lining, and the cells forming reproductive gametes are all highly sensitive and will be damaged by radiation doses which show no detectable effect on cells of the bones, muscles, and nerves. Fortunately, cancer cells are metabolically active and rapidly dividing. Thus, they can be destroyed by radiation which will not injure the surrounding tissue; the differential, however, is not very great and care in measuring the dose administered is necessary to avoid damage to the healthy tissue surrounding the cancer.

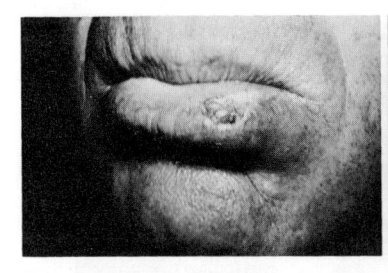

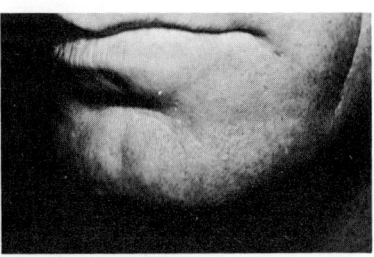

Fig. 17.6. Cancer destruction by radiation treatment. The cancer shown on the lower lip at the left has completely disappeared after several months following X-ray treatment as shown on the right. (Courtesy J. Fred Mullins, University of Texas Medical School.)

Medical Uses of Radiation. In addition to X rays, medical science has found many of the radioactive isotopes valuable for studying internal body structures. Iodine 131 is extremely useful in connection with thyroid disease. Because most of the iodine taken into the body goes to the thyroid gland, this isotope can be used to check iodine uptake by the gland and to locate and treat thyroid cancers. The patient drinks a radioactive iodine "cocktail" and by means of suitable instruments the location of cancerous thyroid tissue can be determined. Radioactive iron is valuable in studies of blood cell abnormalities since it is absorbed by the red blood cells. Phosphorus 32 is of value in leukemia treatment because it is taken up by the bones and can destroy the cancer-like bone marrow that is producing an excess of white blood cells. Cobalt 60 gives a powerful, easily controlled and directed source of gamma rays for treatment of cancer that has advantages over X rays in many cases. In spite of all these benefits, however, radiation can be dangerous and injuries from improper use can occur.

Radiation Injuries. Many of the early X-ray workers suffered serious injuries and painful deaths because they did not understand the dangers of extensive exposures. Thomas Edison, upon

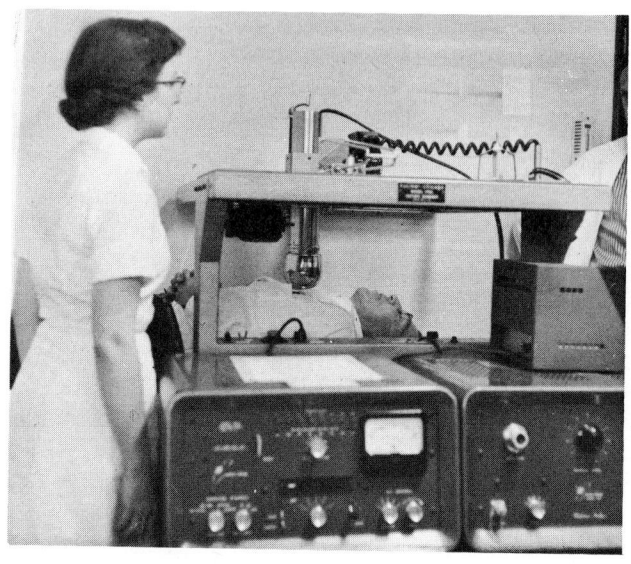

Fig. 17.7. Detection of thyroid cancer with automatic counter. This patient previously drank a radioactive cocktail containing iodine 131. The counter detects areas of the chest where pieces from a thyroid cancer may have migrated and are growing.

hearing of the work of Roentgen, developed an X-ray tube and discovered that the rays would produce a fluorescent image on a chemically-coated screen. This fact was demonstrated at an electrical exposition in New York in 1896, and many people marveled that they could see the bones of their bodies on the screen. Edison's assistant, a Mr. Dally, repeatedly exposed himself in demonstrating the newly discovered marvel and as a result suffered severe

X-ray burns. These grew more serious with the passage of time and he died of radiation injury in 1905.

Other workers lost fingers or entire hands as a result of overexposure, and it would be assumed that these cases would lead to proper respect for these powerful rays, but instead, cases of injury and deaths mounted over the years into the hundreds. It was learned that one effect of heavy exposure was removal of hair due to the damage to the hair follicle cells. A doctor in a small town had a young secretary who complained about the hair on her arms. In order to remove the hair, he gave her extensive treatment with X rays, which caused a serious "burn" of the skin. In time, cancerous growths developed and both arms had to be amputated.

Not deterred by these tragic cases, many X-ray machines were installed in beauty shops during the 1920's and advertised as a wonderful new discovery for removing superfluous hair. One physician inaugurated the "Tricho System" wherein patients were exposed to nearly 500 r every two weeks for a period of about 12 weeks. As the years passed, there were many cases of ulcerations and cancer resulting from this treatment.

In the 1930's shoe stores installed fluorescent machines to show customers the bones of their feet inside the shoes. When some of the shoe salesmen demonstrated too often on their own feet there resulted many cases of foot damage and amputations. In spite of this, some of these machines are still in use today.

Dentists learned that when they used their thumbs to hold X-ray film in their patients' mouths they sometimes developed injuries that resulted in amputation.

The Cumulative Effects of Radiation. Most of these tragic cases came about because the persons involved did not realize that, while there is some repair of acute radiation injury, there is also a certain amount of irreversible damage which is cumulative over one's lifetime. Skin which reddens after heavy exposure may gradually recover. The severe anemia and the reduction in leucocytes may gradually be overcome as new bone marrow cells grow in and replace those destroyed. The ulcerations of the intestine may heal in time. However, all of the exposed tissues have received a certain amount of irreparable damage and therefore have a lowered tolerance to additional radiation. This damage may show up in future years as cancer, cataract of the eye, lowered

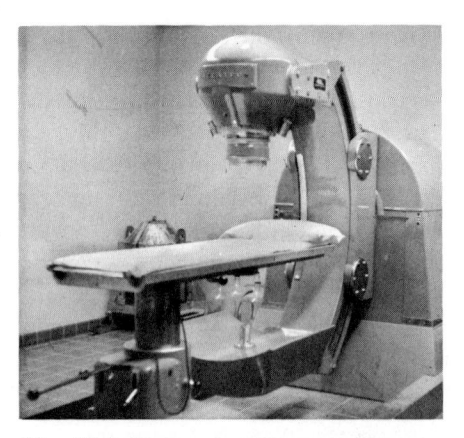

Fig. 17.8. Cancer treatment machine using cobalt 60 radiation. There are several grams of Co^{60} in the head of this machine. When the filters are removed a very powerful beam of gamma rays can be directed onto cancer tissue. The head can be rotated to concentrate the beam on deep-seated cancers. The machine is located at the University of Michigan cancer clinic.

vitality, lowered resistance to disease, and a decreased life expectancy.

Atomic Radiation. It was not long after the discovery of radium that the Curies learned of the injurious effects of its rays. Becquerel was given a sample in a small glass vial which he carried in his pocket for several days. A burn developed at the spot. Pierre Curie then purposely exposed a portion of his arm for ten hours and studied the reddening effects and the later ulceration. He was probably spared death from radiation only because he was killed when run down by a carriage, but his wife continued with the investigation and in time died from the delayed effects of the radium rays which she had absorbed.

Again, the lesson from such experiences was ignored. Radium treatments for arthritis became very popular about 1915 and as late as 1930 people were still drinking and being injected with radium salts. These radioactive materials become lodged in the body and produce a continual radiation. Needless to say, debilitating injuries and deaths from these treatments have been found to be extensive.

The people of today are faced with the critical problem of controlling new sources of radiation. Man has discovered the atom and the hydrogen bomb. These release radioactive elements in great quantities, some of which are spread over the entire surface of the earth. We do not need to review the horrible cases of radiation damage which resulted from the bomb explosions in Japan. There is growing concern among many who are familiar with the situation that continued bomb testing programs, even though the bombs are never used in wartime, can cause radiation damage via the fall-out that settles over the earth. This situation is especially important because of certain isotopes that become concentrated in living tissue. **Strontium 90** (Sr^{90}) for example, is similar to calcium and tends to replace calcium in both plant and animal tissue. Thus, a plant may have many times as much strontium 90 in its tissues as is found in the surrounding soil. A cow eating this plant will further concentrate the isotope in the milk and a child drinking this milk will further concentrate it in his bones. Sr^{90} gives off beta particles and has a half life of about 28 years. Leukemia is known to result from heavy Sr^{90} concentration in the bones and there is some fear expressed that the incidence of this horrible affliction may increase if the Sr^{90} continues to be released in our atmosphere through hydrogen bomb tests.

GENETIC EFFECTS OF HIGH-ENERGY RADIATION

Scientists have learned some of the biological consequences of heavy exposure to high-energy radiation. Horrible as the immediate effects may be, they are thought to be matched and exceeded by the damage to future generations of mankind through mutations induced in reproductive cells. Chromosomal aberrations can also play a part in producing abnormal offspring, but they tend to be self-eliminating and do not present the hazards found in gene mutations. In view of the possible sources of radiation to which man will be exposed, it is important that we have some knowledge of the effect of this radiation in mutation production in man.

The Mutation Doubling Dose. Man's mutation rate is noticeably increased by radiation-induced mutations. A great deal

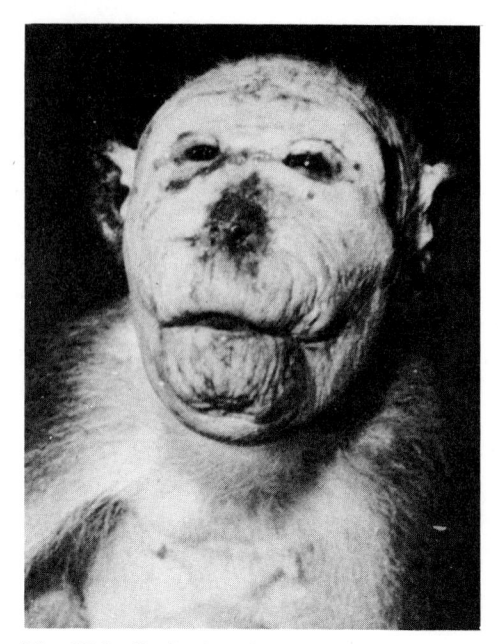

Fig. 17.9. Radiation damage to living tissue. The head of this monkey was exposed three months previously to cobalt 60 and received 2,000 r of gamma rays. (Courtesy A. J. Riopelle.)

of study has gone into determining the number of mutations which will be induced by a given dose of radiation. Leading geneticists are generally agreed that exposure of the gonads to a dose of about 30 to 80 r will produce mutations equal in number to those which occur naturally. This is the so-called doubling dose, and 50 r is used as a good working estimate. It should be realized, however, that this is only an estimate and can be incorrect. In fact, the recent work of Russell and his co-workers at the laboratories at Oak Ridge, Tennessee, indicate that the doubling dose for mice may be as low as 10 r. We will use, however, the 50 r figure for the following discussion, knowing that this figure may be adjusted as more information is obtained.

It will help us to evaluate the damage which might be done by a doubling dose if we know the amount of damage now done by natural mutations. About 4 per cent of the children born in the

United States have clearly defined defects (for example, malformed body organs, blood or hormone defects, impaired sight or hearing, epilepsy, mental abnormalities, neuromuscular defects, etc.). About half of these defects are due to embryological accidents, birth injuries, and other factors not related to heredity. The remaining 2 per cent appear to have a genetic origin. This figure tends to remain constant because of the rate of elimination of the abnormalities from the population as a result of the afflicted individual's inability to survive or reproduce. If those living in the United States were to receive radiation sufficient to double this rate in each generation, there would be a gradual increase in these abnormalities until they reached 4 per cent.

This estimate becomes more vivid if we use specific figures. There will be about 100 million children born to the people living in the United States today. About two million of these will show abnormalities due to harmful mutations which occurred in present or past generations. If a doubling dose of radiation is received, this number will be about 10 per cent greater—about 200,000 more abnormalities will occur. If the number of mutations were to be doubled in each succeeding generation, eventually there would be a doubling of abnormalities, at which point a stabilization would be reached with eliminations balancing mutations.

Effect of Intensity of Radiation Dose. The *Drosophila* studies have indicated that the number of mutations resulting from radiation are in direct proportion to the amount of radiation given. A dose of 800 r causes twice the number of mutations that are found after a dose of 400 r, etc. The kinds of mutations are the same—they are no more severe in their effects after heavy radiation than after smaller doses—just more mutations occur with the heavy dose. It is difficult to carry on experiments with low doses because large numbers of offspring must be examined in order to determine the small percentage increase of mutations over those that occur naturally. This information is of great importance to man because people today are likely to be faced with the problem of widespread exposure to radiation of comparatively low dosage. If extrapolation is used, we would assume that one r will produce one fiftieth as many mutations as 50 r. Thus, a million people, each receiving one r, would receive the same total number of mutations as 20,000 people receiving 50 r each. This method assumes a

250   RADIATION HAZARDS IN AN ATOMIC AGE

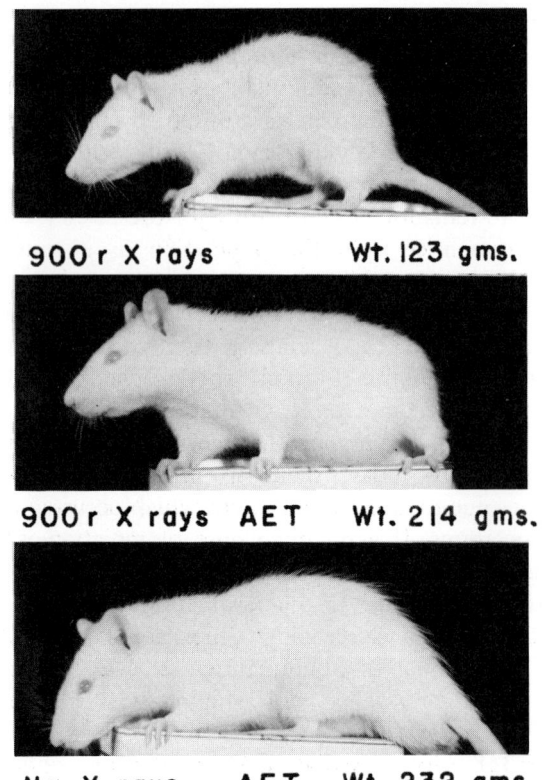

900 r X rays   Wt. 123 gms.

900 r X rays AET   Wt. 214 gms.

No X rays   AET   Wt. 232 gms.

Fig. 17.10. *Chemical protection from radiation damage. The chemical* AET *gives some protection against radiation damage if it is given before exposure as shown by these pictures of typical rats from groups treated as indicated.*

linear response with no threshold. Some scientists have questioned the no-threshold concept, and experiments on mice conducted at the Oak Ridge laboratories have tried to solve this problem. Thus far, some startling facts have been uncovered. The results obtained from heavy gonadal dose (1,000 r) revealed a highly significant departure from linearity if 0 to 600 r results were extrapolated. The number is lower than would be expected. This situation might occur because of the elimination through death of those cells most susceptible to mutation. If this principle were extended to the low doses, we might find the mutation rate at low dosage to be higher than any estimate based on a linear extrapolation between the 600 and 0 r rates.

Effect of Distribution of Radiation Dosage. Another important question involving the future of mankind concerns the distribution of the radiation. Is there a reversible genetic damage which to some extent can be repaired between exposures if the radiation is spread out in time instead of being delivered all at once? *Drosophila* studies indicate that there is no such recovery. The number of mutations observed after 2,000 r of X rays given at intervals over a week's time is about the same as the number found in the descendants of flies receiving 2,000 r in one dose. The present studies on mice at the Oak Ridge laboratories indicate that the number of mutations produced by chronic irradiation (extended over a period of time) is at least as high as the number found after a single radiation dose of the same total intensity. Hence, we would assume that there is no threshold and no recovery since, if we are to err, it must be on the side of caution.

RADIATION TOLERANCE DOSES

The toleration to whole body radiation is often expressed in terms of LD 50 (that is, the amount of radiation required to kill 50 per cent of the exposed individuals). In man this figure is set at about 450 r whereas 600 r will kill nearly all exposed individuals. The LD 50 figure varies with different species of organisms. The following table indicates this variability.

LD 50 for Different Organisms

Organism	Radiation in Roentgens
Dog	325
Man	450
Mouse	530
Rat	850
Drosophila	100,000
Bacteria	20,000 to 50,000

The above table reveals that man is one of the more susceptible organisms. This situation is likewise true for mutation induction. In *Drosophila* the amount of radiation required to double the mutation rate is about 400 r, but in man it is around 50 r. Of course, isolated parts of the body can stand a much higher dosage. For example, between 4,000 and 5,000 r is frequently used to kill cancer tissue, but the radiation is carefully directed so as to expose a minimum amount of healthy tissue. If the entire head is exposed, with the rest of the body shielded, it requires about 2,000 r to cause death. The abdominal region can stand a dosage somewhere between 3,000 and 5,000 r, but this does not mean that no damage will occur at lower dosages. Studies made by Dr. Michael Bender of the Johns Hopkins University revealed that doses of 50 r would cause clearly visible chromosome breaks in about 20 per cent of the cells in a tissue culture of human cells. Cataract of the eye, delayed cancer, blood disorders, etc. can be caused by much lower dosages.

In addition, embryos and young growing organisms are much more susceptible to radiation damage than adults. In Hiroshima many pregnant women had natural abortions after exposure to the heavy radiation. The embryos had been killed by radiation which did not kill the adult women. Since human embryos are especially sensitive during the formative stages of early development, all radiological examinations and treatments of the abdominal region should be avoided, if at all possible, at this time.

EXPOSURES FROM MEDICAL USE OF X RAYS

In the light of the information just given, we might illustrate the amount of radiation received in medical practice. It should be understood that these figures are quite variable, depending upon the type of machine used and the technique of the operator. A combination of an ancient machine and an operator not trained in modern methods of reducing exposure to the patient may produce an exposure more than several times that indicated.

Chest. Using chest-sized film: .05 to .5 r.
Mobile X-ray unit: 1 r.
Chest fluoroscope: ranges as high as 130 r.
Gastrointestinal Examination. About 5 films and 6 quick fluoroscope spot checks: 15 r.
Gall Bladder Examination. About 5 films: 6.5 r.
Teeth. Single dental picture: 2 r.
Complete mouth: 20 r to 150 r.

Warts. 1,000 to 1,500 r.
Cancer. 4,000 to 5,000 r.
Acne. 700 to 800 r of soft X rays delivered to the skin.

In most instances the exposure to the gonads need be very small if proper precautions are taken to shield those areas of the body not needing X-ray treatment.

THE THIRTY-YEAR GONADAL DOSE

For purposes of evaluating the possible genetic damage which might be caused by the agents of radiation to which man is exposed today, scientists have worked out the average amount that will reach the gonads of an individual from the time of his birth until his thirtieth birthday. Thirty years is chosen because this figure presents the average age of parents at the time of conception of their children. A child can receive a mutant gene induced at any time during the past life of its parents. The sources of radiation and their estimated quantity are expressed below.

SOURCE OF RADIATION	ROENTGENS RECEIVED BY GONADS IN THIRTY YEARS
Background	3.1
Medical X rays	4.6
Fall-out (if bomb testing continues at present rate.)	.1

The radioactive fall-out is low in comparison to other sources of radiation to which man is exposed, but because of its wide distribution and the large number of people exposed, there are many who feel that it can be significant. Scientists could better evaluate the possible effect of the radiation if they possessed more accurate information on the genetic effects of radiation on man. In Japan there were masses of people in the two cities that received heavy radiation from the atom bombs. Many people in the affected cities now have children. Drs. Neel and Shull of the University of Michigan, in co-operation with Japanese investigators, have made extensive studies of these people and their offspring.

THE ATOM BOMB STUDIES IN JAPAN

A group of American and Japanese physicians were especially trained for this study. Over a six-year period an attempt was made to study every child descending from parents who had been exposed to radiation from the atom bomb. Also studied was a control group consisting of a large number from a city that had received no radiation. The parents were classified according to their distance from the explosion and to the amount of protection they had had. These figures yielded an approximation of the exposure they received. The children were classified according to birth weight, progress of growth and development, infant and fetal mortality, congenital malformations, and sex ratio. The results up to this time have failed to show a significant increase in damage to the children, even those of parents who received a dosage of about 200 r each because of their close proximity to the bomb. It is easy to jump to the conclusion from these results that we have little to fear from radiation as far as genetic effects are concerned. Drs. Neel and Shull, directors of the study, would be the first to warn against such an unwarranted assumption. Knowing what we do about the nature of mutations—recessives, detrimentals, and lethals—we would expect a small increase in easily detected abnormalities. An increase of 10 per cent would require many figures in order to detect them at a significant level. With respect to the sex ratio, however, we have a different story. Recessive lethal mutations which appear on the X-chromosome of a woman's reproductive cell could cause the death of a son. These deaths could be detected through the lowered number of males in the children of radiated women. Studies of the sex ratio in cases where the mothers received heavy radiation from the bombs have revealed a significant reduction of males. No such reduction was found in the other cases even though in some instances the father received heavy radiation. Such results would be expected if the radiation increased the number of recessive lethals on the X-chromosome. This is the first concrete evidence we have that there is danger to future generations of man from radiation produced by nuclear weapons.

REDUCTION OF RADIATION HAZARDS

In the light of both the biological and genetic hazards from radiation it is important that exposures are kept to a minimum. It is almost certain that the future will see an ever-increasing use of radioactive materials in industry, in research, in medicine, and as a source of power. The benefits to be derived from such uses are too great to justify the recommendation that we stop using all sources of radioactivity. Automobiles take a frightful toll of human life each year, yet no one suggests that we go back to the horse-and-buggy mode of transportation. The great advances of medical science are saving many people with inherited defects, thus allowing them to reproduce. This situation will produce more such defective persons in the future generations, yet no one would suggest that we abandon our medical achievements. All gains are bought with a price and we must learn to reduce the price we pay for these gains by application of principles which will reduce the harmful effects. Likewise, man-made radiation is here to stay and we must learn to live with it, doing all we can to reduce any harmful effects that it may have on present and future generations.

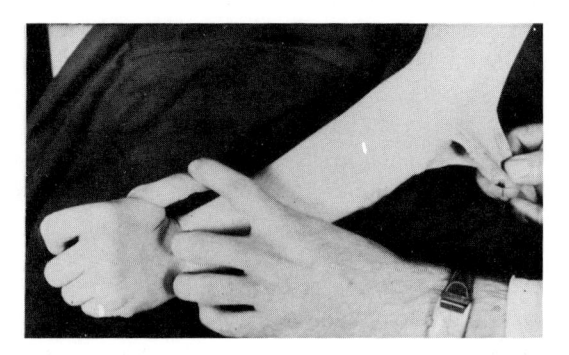

Fig. 17.11. A gross human abnormality of the kind which can appear as a result of a gene mutation. This is Ehlers-Danlos syndrome which is characterized by extreme flexibility of the joints, looseness and elasticity of the skin, and fragility of the capillaries of the skin. (Courtesy T. Eleigelman and Sture M. Johnson.)

Committees Concerned with Radiation. It was in the light of the above reasoning that a committee of distinguished scientists was appointed by the National Academy of Science to study the problem and make recommendations to reduce radiation hazards. The committee included physicians, radiologists, physicists, and geneticists—workers in those fields related to radiation. Their recommendations are summarized below. (Similar committees in Great Britain and elsewhere have made like recommendations.)

Recommendations to Reduce Radiation Hazards. (1) Records should be kept for every individual, listing the total lifetime accumulation of exposure to radiation in each part of his body. Each person could keep a copy of his record and every time any sort of exposure to high-energy radiation occurs (such as an X-ray picture), the number of roentgens received and the part of the body receiving them would be recorded. This procedure would enable a physician to avoid excessive exposure of a patient who had already received previous extensive exposure to the same body parts.

(2) The medical use of X rays should be reduced as much as is consistent with medical necessity, and particular care should be taken to shield the reproductive cells from radiation. When other means of diagnosis or treatment are available, they might well be substituted for extensive use of radiation. In addition, each X-ray technician should know the output of his machine in terms of roentgens per minute at a specified distance and use the smallest exposure necessary to obtain the desired results. In one survey of a large number of machines it was found that few users knew the output of their machines and therefore individual technician's exposures varied as much as fortyfold in the number of roentgens used.

(3) The average exposure of the population's reproductive cells above the natural background radiation from conception to age 30 should be limited to 10 r. Of course, some would need more, but this is the recommended average maximum.

(4) No one individual should receive a total accumulated dose to the reproductive cells of more than 50 roentgens up to age 30, and not more than an additional 50 up to age 40. (About one half of the children in the United States are born to parents under 30 and nine tenths to parents under 40.)

18: HEREDITY AND ENVIRONMENT

At the beginning of the present century, when the particulate concept of inheritance was formulated after the rediscovery of Mendelism it was customary to think of all genes as being expressed with unvarying precision in accordance with certain predictable ratios. There was a tendency to think of all characteristics as due either to heredity or environment and arguments often arose as to whether some particular characteristic was due to one or the other. Modern genetic study, however, has shown that there is no clear-cut line of distinction between inherited and environmentally-induced characteristics. As a matter of fact, it is a blend of the two which results in the final expression of most characteristics. In this chapter we will explore this topic more fully.

EXPRESSIVITY

There are many cases (in different forms of life) where organisms with the same genotype with respect to a particular gene show a variation in the degree of expression of the characteristic involved. *Drosophila* furnishes us with a good example of this condition.

Vestigial Wings in Drosophila. A recessive autosomal gene, *vg*, causes the wings to be mere stumps provided that the flies are grown at a temperature no higher than 72 degrees. An 8-degree elevation of the temperature will cause many of the flies to have wings considerably larger than the vestigial size and another 8-degree elevation will cause them to be so large that they extend beyond the abdomen (see Fig. 18.1). Hence, it is evident that the temperature, in some way, has an influence on the degree of expressivity of this gene. In one environment it has one phenotype; in another environment the phenotype is so different that the

Fig. 18.1. Variation in expressivity in the wings of Drosophila. The flies in these photographs are all homozygous for the recessive gene for vestigial wings. Temperature, however, influences the size of the wings for such flies. The fly at the left was raised at normal room temperature of about 72 degrees. This is the expression of the gene usually observed. The second fly was raised at 80 degrees and the third at 88 degrees.

characteristic, if it were not investigated, might be taken for an influence of a different gene.

The Himalayan Coat Pattern in Rabbits. In rabbits, there is a recessive gene, *h*, which produces the Himalayan coat pattern. Usually such rabbits are white over most of the body, but the feet, ears, tail, and tip of the nose are black. The gene may show variation in expressivity, however, if the temperature of the skin is subjected to unusual chilling. For instance, if a naked, newborn rabbit is exposed to a temperature of 52 degrees for a short time and then returned to its mother, the skin will grow out black all over the body. Likewise, a foot will develop white hair if it is bandaged in such a way as to keep it warmer than normal. These examples reveal that there must be two factors acting in the production of the coat pattern, one genetic and the other environmental.

The unusual results described above might be explained by theorizing that the gene for this coat pattern produces an enzyme which is necessary for the formation of black pigment, but that (1) at temperatures above 92 degrees it will not produce the enzyme or (2) the enzyme will not function. The body heat keeps the skin above this critical level over most of the animal; the extremities, however, are readily chilled and thus the skin drops below this temperature. Hence, there occurs the development of the black hair. If the fur is plucked from the back of a Himalayan

EXPRESSIVITY 259

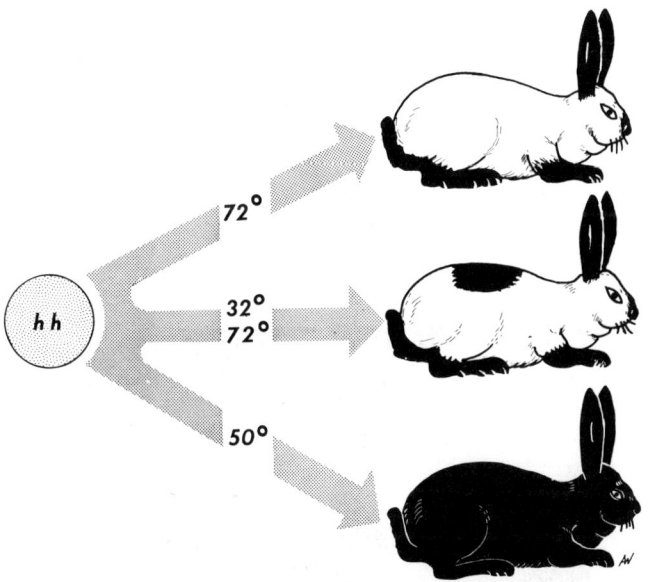

Fig. 18.2. *Effect of temperature on expressivity of the gene h in rabbits. All of these rabbits have the same genotype. The one at the top was raised at 72 degree temperature and shows the typical markings of the Himalayan rabbit. The second rabbit had an ice pack applied to his back for a time after hair had been plucked from this region. The black rabbit was chilled when newborn before hair had begun to grow. The black hair grows only when the temperature drops below a critical minimum level. (From Winchester, Heredity and Your Life, Dover Press.)*

rabbit and an ice pack is placed on this region for a time, the hair which grows back will be black. The same results could be explained if it were assumed that the genes for the Himalayan pattern produce an inhibitor which prevents black pigment from forming in rabbits which would otherwise would be black all over. At temperatures below 92 degrees, however, this theoretical inhibitor cannot function and thus black pigment is formed.

The two examples presented by *Drosophila* and rabbits demonstrate how environment from an external source can alter the expressivity of genes. There are other cases where the environments appear to be identical, and yet there occurs variation in expressivity. Such variation must come from some internal factors, and in most cases these factors are other genes. There may be one major gene which brings about some particular phenotypic effect, but modifying genes may affect the phenotype, thus altering the expressivity of the major gene.

The Eyeless Gene in Drosophila. There is a recessive gene located on the fourth chromosome of *Drosophila* which produces what is called the eyeless phenotype. When you see a group of flies homozygous for this gene, however, you will find quite a

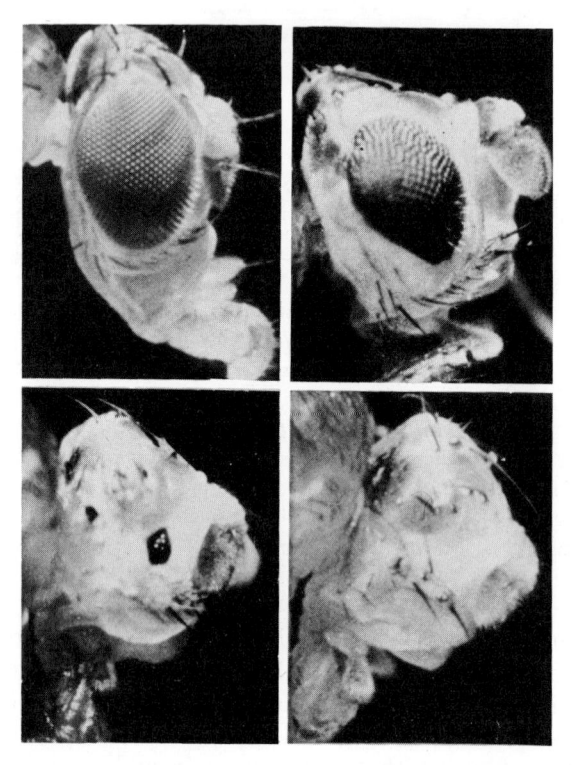

Fig. 18.3. Variation in expressivity of the eyeless gene in Drosophila. The normal eye is shown in the upper left, the other three flies are all homozygous for the gene for eyeless and were all raised in the same vial. The variation in the size of eyes from nearly normal to no eyes at all is apparently due to other genes which modify the expression of the eyeless gene.

variety of eye sizes represented. Some will have eyes almost as large as normal, whereas in some the eyes may be almost or completely lacking. The majority of cases will range between these two extremes. The variations in eye size will be found among flies raised in the same vial. It appears that there must be other genes at work which modify the effect of the gene for eyelessness. Through selection for both extremes of eye sizes in an eyeless stock, we can accumulate a group of modifying genes which will produce relatively large eyes in one stock and very small eyes in another.

Variable Expressivity of Human Genes. There are many human genes which have variable expressivity, for example, the dominant gene for *blue sclera* of the eyes. The sclera is a part of the eye that is normally white. Some individuals carrying this gene may have a very pale blue sclera while in others the expressivity may vary so that there are expressed various shades of blue—even a blue so dark that the "whites" of the eyes look black. It appears that modifying genes account for this great variability. Unfortunately, this gene does more than affect the color of the sclera—it also may affect the bones. About three fourths of those with blue sclera also have fragile bones which break easily. The degree of fragility is quite variable. In some cases a bone will break if caught on the sheet while a person is turning over in bed. In other cases breaks require greater strain, but

Fig. 18.4. Variable expressivity of a dominant gene in man. These men are father and son and both carry a dominant gene affecting the arms. In one there is a shortening of the arms and some deformity of the hands. In the other the arms are reduced to mere flipper-like stumps. (Courtesy Karl A. Stiles, Michigan State University.)

in all cases the bones break more readily than in unafflicted persons. In some the condition is outgrown when adulthood is reached; in others it may persist into old age. Thus, there is a considerable variation in the degree of expressivity of this gene.

In the condition known as *brachyphalangy* a shortening of the second bone of the fingers occurs. It is brought about by a dominant gene which is considered to be lethal when homozygous. In the heterozygous condition there exists considerable variation in the degree of shortening of the bone. In some the bone is so short it appears to be absent and thus the fingers seem to have only two bones—like thumbs. This disorder varies from slight shortening to the severe shortening already discussed.

PENETRANCE

In some cases the expressivity of a gene is carried so far that individuals do not show any phenotypic evidence of the gene. The human gene for blue sclera again can be used to illustrate this point. In about one tenth of the people who carry the gene there is no detectable expression of it; yet about one half of their children will exhibit some sign of blue sclera. Penetrance is usually expressed in terms of per cent of occurrence. In this case the gene would have 90 per cent penetrance—it shows itself in 90 per cent of those who carry the dominant gene.

Nicked Wings in Drosophila. The recessive gene for nicked wings in *Drosophila* is a good example of a gene with a low penetrance. The phenotypic expression of this gene is a small nick in the outer edge of the wings. The penetrance of this gene is only about 3 per cent; i.e., about three out of every one hundred homozygous flies express the nicked wings. If we cross two flies with nicked wings we find the trait showing in about 3 per cent of their offspring, and if we cross flies from the same culture that do not have nicked wings we find the same percentage of nicked wings expressed in their offspring. Hence, we know that the genotype of the two sets of parents was the same with respect to this one gene, yet one set carried the gene without expressing it.

Tremor in Chickens. A recessive gene in chickens may cause them to shake almost continuously. Hutt and Child studied 112 homozygous chickens and found that only 39 showed a detectable tremor. The others appeared perfectly normal, yet the descendants of the seemingly normal chickens also showed tremor in about the same ratio. In addition, the chickens that show the tremor reveal a variation in expressivity. Some shake so violently that they have great difficulty in taking food, while others have a tremor so slight that it is barely perceptible. Hence, the gene is one with a penetrance of about 35 per cent and a great variation in expressivity.

Human Genes with Reduced Penetrance. Many human genes have a reduced penetrance and it is this factor that complicates the study of inheritance in relation to various diseases and abnormalities. The gene for *multiple exostoses* causes abnormal growths on the bones, but this gene has a approximate penetrance of only 60 per cent. Thus, even though it is a dominant gene, the condition can be inherited and expressed by a child when neither parent reveals the condition.

In many cases where there exists reduced penetrance, scientists have found that delicate tests can reveal some degree of expression of the gene in those who apparently are not influenced by it. *Gout* is a disease characterized by an abnormally high level of uric acid in the blood; the acid may crystallize out in the joints thereby causing the painful swelling which is the primary symptom of this disease. Judged on the basis of the appearance of gout in the joints, this dominant gene has low penetrance (about 10 per cent in men), but a chemical analysis of the blood shows unusually high uric acid concentration in 100 per cent of those who carry the gene. Thus, gout is actually a case of variable expressivity rather than absolute penetrance.

PHENOCOPIES

Studies of heredity and environment are complicated by the fact that there may be environmentally induced characteristics which are exact copies of characteristics normally produced by genes. The term phenocopy has been applied to such cases.

Phenocopies in Drosophila. Richard Goldschmidt, of the University of California, found that many mutant phenotypes of *Drosophila* could be induced environmentally by heat shocks. For example, he found that if larvae between four and one half to five and one half days old were exposed to a warm temperature of 35 degrees centigrade from twelve to twenty-four hours, about

70 per cent of the adults that descended from these larvae would have scalloped wings. This characteristic also results when the flies are homozygous for a certain gene regardless of temperature. The heat-induced and the gene-induced phenotypes are indistinguishable. When the larvae were kept until they were seven days old and then subjected to the same heat shock, the adults emerged with miniature wings in about 40 per cent of the cases. Miniature wings are also produced by a recessive sex-linked gene.

On the basis of these results, and from other phenocopies he obtained, Goldschmidt postulated that mutant genes produce their effect by altering metabolism at specific stages of embryology. The body structures which are at a critical stage of development at this point will be altered and, therefore, the adult reveals a specific phenotypic alteration. Likewise, scientists can duplicate the effect of mutant genes by an environmental alteration of the metabolism at the same critical time during embryonic development.

Harelip in Mice. In certain strains of mice, harelip appears in about one fourth of the offspring no matter how ideal the environment of the mother may be. In other strains the abnormality under normal environmental conditions is practically unknown. This fact establishes beyond doubt that there is some hereditary background for the harelip abnormality. It is possible, however, to obtain a high incidence of harelip from the second strain by using certain treatments on the pregnant females at a time when the embryos within their bodies are at a specific stage of development. The treatments can be quite varied; in fact, it does not matter what treatment is used, for it is the time of application that is the important factor. Injections of cortisone, feeding a diet deficient in certain vitamins, subjecting the females to an atmosphere of low oxygen concentration, and removal of some of the amniotic fluid around the embryos are examples of the treatments that have been tried. All the treatments interfere with normal metabolism at a time when the upper lips of the embryos are at the stage of fusion. The lips will not fuse properly and the time for fusion passes. Thereafter, even if normal metabolism is restored, the embryos are unable to accomplish the fusion; thus, harelip results. In the strains where harelip occurs without the environmentally induced alterations of metabolism, we can assume that the genes accomplish the same thing.

Phenocopy in Man. There is evidence that a relationship similar to the phenocopy in *Drosophila* exists in man. Some physicians advise women in the early weeks of pregnancy to avoid extended exposure to conditions of low oxygen concentration such as might be encountered on visits to high-altitude locations or even high-altitude flying. In addition, the administration of strong drugs, radiation treatments, or any other agents influencing the developing embryo can conceivably produce abnormalities similar to those abnormalities caused by genes. Most of the body organs develop during the first six weeks of embryonic development. We know that if a woman has German measles during this early embryonic stage the child may be born blind due to cataract of the eyes. A dominant gene can produce the same condition even though there occurs no shock to the embryo. There is also evidence that harelip in men follows the same pattern as harelip in mice (see above). Other abnormalities have been found to occur according to the time of embryonic life when the shock is received.

TWIN STUDIES

Studies of twins and other multiple births provide geneticists with an excellent opportunity to determine the effects of heredity and environment in man. Identical twins, having descended from one fertilized egg, have identical genes; therefore, any differences which they manifest are obviously due to environmental agents. Fraternal twins, on the other hand, originate as two separate fertilized eggs and will differ in many of their genes. Thus, differences which fraternal twins manifest can be due to heredity, environment, or a combination of both. Fortunately, from the standpoint of genetic studies there are a few cases in which identical twins have been separated shortly after birth and raised in different environments. If we assume that twins of the same sex receive about the same environment when raised in the same household, we have three sets of circumstances for examination. First, we have children with the same heredity and approximately the same environment—the identical twins raised together. Second, we have children with some differences in heredity and approximately the same environment—the fraternal twins of the same sex raised together. Finally, we have children with the same heredity and different environments—identical twins raised apart. Comparisons

Fig. 18.5. Identical twins. These young ladies started life as a single fertilized egg which separated into two parts during early embryology. They have identical heredity and any of the differences which they exhibit are certain to be due to environment.

of these three groups have provided insight into the influence of heredity and environment on different characteristics.

Results Obtained from Twin Studies. The first extensive investigation of characteristics of different types of twins was done by H. H. Newman at the University of Chicago. Since that time literally thousands of twins have been studied with respect to those characteristics which might have some basis in heredity. The results of some of these investigations are tabulated below. In this table the results are given in terms of differences. For comparison a fourth group is included. These are the "sibs"—a term denoting brothers or sisters who are not twins. This group is of different ages, but their gene similarity is the same as for fraternal twins—when allowance is made for the age differences we would expect them to show about the same variability as fraternal twins. The sibs and fraternal twins considered in this study are always those of the same sex. Sex makes such a difference in characteristics and in environment that it would yield erroneous results if we included fraternal twins or sibs of opposite sexes. Identical twins, of course, are always of the same sex.

The results point out that heredity plays a major role in the physical characteristics such as, for example, body height. Fraternal twins of the same sex show a distinct difference from iden-

DIFFERENCES BETWEEN TWINS AND SIBS

	Identical	Fraternal	Sibs	Identical Reared Apart
Body height (difference in cm.)	1.7	4.4	4.5	1.8
Body weight (difference in lbs.)	4.1	10.0	10.4	9.9
Total finger ridges	5.9	22.3	—	—
Age of first menstruation (difference in mo.)	2.8	12.0	12.9	—
I.Q. (Binet) (difference in points)	5.9	9.9	9.8	8.2
Criminal record (difference in per cent of concordance)	32.0	72.0	—	—

tical twins even though the environmental agents seem to be the same for the two types of twins. Even when identical twins are reared apart they still show a similarity in stature which is almost the same as among those reared together. Environment seems to play a greater role in the determination of the I.Q. scores—the identical twins reared apart reveal a considerable variation from those reared together. Still, the fact that fraternals are more variable than identicals reared together shows that heredity does play an important role.

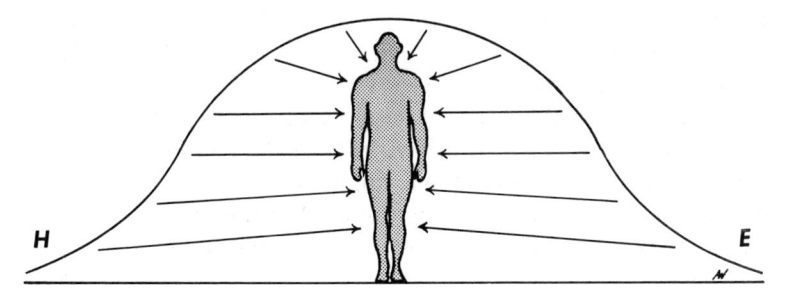

Fig. 18.6. Heredity and environment. Most characteristics represent a blend of these two important factors, but a few are due almost solely to heredity and a few almost solely to environment.

Concordance. Many studies of twins are made on the basis of concordance, or the percentage of similarity. A study of police records revealed that a certain number of individuals who had committed major crimes possessed a twin brother or sister. A study was then made to determine whether the twin also had a criminal record and whether the twins involved were identical or fraternal. In 68 per cent of the cases in which the twins were identical, both were found to have a criminal record. In the cases in which the twin was fraternal and of the same sex, only 28 per cent were found to have a criminal record. A similar study in Germany gave concordances of 68 and 38, respectively.

In studies of this nature it is important to know whether the twins involved are identical or fraternal. Fraternal twins often look much alike when they are of the same sex just as siblings born at different times may look alike. Studies of fingerprints are a reliable method of determination in those cases in which doubt may exist. The finger friction ridges may differ from one hand to another in the same individual, but when the heredity is the same the ridges are much alike and the gene disparities found among fraternal twins usually constitute sufficient evidence in establishing clear-cut identification. There are rare cases, however, in which some doubt may remain, and it is then that the investigator usually resorts to skin grafting. It is known that skin ordinarily cannot be grafted from one person to another. The differences in the antigens of the protein component of the skin are so varied because of the differences in heredity that the body will not accept a foreign skin graft. In the case of identical twins, however, the genes are the same—skin can be grafted and will take as easily as a graft from a person's own body. Thus, an attempt may be made to graft a tiny bit of skin as an ultimate check on the nature of a pair of twins. The physiological characteristics are a better indication than the physical characteristics. There have even been some cases in which an entire kidney has been grafted from a person to his identical twin. If there were any gene disparities such grafts would be impossible because of the antigenic difference which would affect the organ and the tissue into which it is grafted.

Environment and Heredity in Disease. One of the most valuable pieces of information stemming from the twin studies has been the effect of heredity on various human diseases and other

Fig. 18.7. Heredity and environment in rickets. Both of these rats have been fed on the same diet, yet the one on the right has developed rickets while the one on the left is normal. The gene differences account for the differences in the amount of vitamin D required to prevent rickets.

afflictions. The results of some of these studies is summarized in the following table with the similarity expressed in terms of concordance.

TWIN CONCORDANCE WITH RESPECT TO DISEASE
AND BODY ABNORMALITIES

Characteristic	Identical	Fraternal
Harelip	33	5
Mongolism	89	6
Mental retardation	97	37
Schizophrenia	86	15
Cancer	61	44
Site of cancer (when both have cancer)	95	58
Measles	95	87
Tuberculosis	87	25
Diabetes melitus	84	37
Rickets	88	22

These studies reveal that even the germ diseases, which one might think would be conditioned primarily by environment, in some cases are related to heredity. Certainly the organic diseases, such as cancer and diabetes, have an important relationship to heredity, but here again some environmental influence can be noted. Rickets is particularly interesting in this sense because it is commonly believed to be caused by a deficiency of vitamin D in the diet. Yet these results reveal the strong influence of heredity in the disease. This statement is borne out by studies on rats. Some rats acquire rickets when fed on the same diet as others who remain normal. By selection it is possible to establish races of rickets-resistant rats that are able to produce normal bone growth on low vitamin D diets, whereas a selected rickets-susceptible race will develop the disease while receiving several times as much vitamin D. We still recognize vitamin D deficiency as the primary cause of rickets, but heredity can play a part concerning the amount of the vitamin necessary for protection.

The more we study heredity and environment the more we realize that the two agents each play a vital, yet separate, role in the development of the individual—few characteristics can be assigned solely to one or the other.

INDEX

INDEX

Aberrations, chromosome, 156–169
Acquired characteristics, 18–19
Albinism, in man, 10–11, 66, 72, 204, 219
  in rabbits, 115
Alkaptonuria, 206
Alleles, 46–48
  multiple, 114–116
Alpha particles, 236
Amaurotic idiocy, 225
Anaphase, 26, 27–28
Ancon sheep, 213, 217
Aneuploids, 160–162
Aniridia, 225
Antibodies, in blood, 128–130
Antigens, in blood, 128–130, 133–139
Aristotle, 14
Aster, 25
Ataxia, Friedreich's, 52
Atom, 234–236
Atom bomb, 247
  studies in Japan, 253–254
Autosomes, 87, 170–175

Bacteria, transduction in, 182
Bacteria, transformation in, 179–181
Bacteriophage, 181–183
Baldness, in man, 110–111
Bar eye, in *Drosophila*, 106–107, 167–169
Bateson, W., 22, 141
Becquerel, Henri, 234, 246
Benoit, Jacques, 184
Benzer, Seymour, 188
Beta particles, 236
Binomial probability, 74–76
Biochemical genetics, 10–11
Blakeslee, A. F., 161
Blood groups, human, 116, 126–139
Blood transfusions, 128–129, 135
Blood typing, 129

Bonnet, Charles, 17
Brachyphalangy, 54–55, 262
Breeding, experimental, 4–7
Bridges, C. B., 171
Bulldog calf, 53–54

Cancer, 217
  radiation as cause of, 242–246
  treatment of, 243–244, 246
Centriole, 25
Centromere, 25
Centrosome, 25
Cerebral sclerosis, 107
Checkerboard method, 60
Chiasmata, 142–143
Chi-square, 79–82
Chlamydomonas, 192
Chondrodystrophic dwarfism, 224–225
Chromatids, 27
Chromonemata, 25
Chromosomes, 8, 9, 23–25
  aberrations in, 156–169, 242
  attached X-, 160, 219–220
  deletion in, 156–158, 166–167
  duplication in, 23–24, 157–158
  human, 9, 24, 163
  inversion in, 158, 159, 167
  mapping of, 148–152
  number of, 24–25, 142
  sex determination by, 87–91, 170–178
  translocation of, 159–160, 167
Cistron, 188
ClB method, 221, 223, 227
Codon, 209
Coincidence, in crossing over, 147–148
Coincident happenings, 71–74
Colchicine, 163
Color blindness, 104–105, 225

273

INDEX

Concordance in twins, 268
Cosmic rays, 237
Counseling, genetic, 3–4
Crossing over, 142–148
Curie, Marie, 234, 246
Curie, Pierre, 234, 246
Cytogenetics, 22
Cytology, 8–10
Cytoplasm, 36, 188–192
Cytoplasmic inheritance, 188

Darwin, Charles, 19–20
Datura; see Jimson weed
Deaf-mutism, 65–66
Deficiency, chomosome; see Deletion
Deletion, chromosome, 156–158, 166–167
Delphinium, 217
Deoxyribonucleic acid; see DNA
DeVries, Hugo, 20–21, 213
Dexter cattle, 54
Dihybrid cross, 59–64
  diagramming, 60–61
  other hybrid crosses, 61–62
  Mendel's experiments with, 59–60
  modified ratios in, 62–63
  short-cut method, 66–67
Diploid chomosome number, 31
Disease, environment and, 268–270
  heredity and, 2–3, 268–270
DNA, 166, 179–187
Dobzhansky, T., 172
Dominance, 42–43
  incomplete, 51–52
Down's syndrome, 178
Drosophila, use in genetic studies, 6–7
Duplication, chromosome, 158, 167–169

Ear lobes, inheritance of, 8–9
Edison, Thomas, 244
Ehlers-Danlos syndrome, 256
Electromagnetic radiation, 236
Empedocles, 13
Encasement theory, 17
Enzymes, and genes, 198–200
  in man, 203–206
  in *Neurospora*, 200–203

Environment, and heredity, 257–270
Epidermolysis bullosa, 107
Epigenesis, 17
Epiloia, 225
Epistasis, 63–66
Erythroblastosis, 134
Euchromatin, 164
Expressivity of genes, 257–262
Eyes, blue sclera of, 261–262
  color of, in man, 73

Fertilization, 36
Fisher, R. A., 137
Four-o'clock plant, 189
Fruit flies; see *Drosophila*

Gamma rays, 228, 236–237
Geiger-Müller tube, 239–240
Gene linkage, 141–152
Genes, 22, 23
  action of, 193–212
  DNA as substance of, 179–187
  duplication in, 23
  and enzymes, 198–206
  expressivity of, 257–262
  intermediate, 62–63
  lethal, 52–55
  location of, 166–167
  multiple, 114, 116–121
  mutation of, 213–232
  structure of, 179–192
  symbols for, 44
Genetic code, 209
Genetics, definition of, 1–4
Genotype, 45–46, 50–51
German measles, 265
Germ plasm theory, 19–20
Glass, Bentley, 85
Goldschmidt, Richard, 174, 187, 263
Gout, 206, 263
Graff, Regnier de, 15–16
Guinea pigs, 60
Gynandromorphs, 95–96
Gypsy moth, 174

Habrobracon, 173–174, 223
Haldane, J. B. S., 226
Half life, 237–239

Haploid chromosome number, 31
Haploid organisms, 191
Hardy-Weinberg principle, 230–232
Harelip, in mice, 264–265
Harvey, William, 14
Height, in man, 117, 122
Hemizygosity, 102
Hemoglobin, human, 193–198
Hemolytic reaction to drugs, 198
Hemophilia, 102–104, 225
Hemorrhagic diathesis, 107
Heredity, and environment, 257–270
Hermaphrodites, 94
Heterozygosity, 46
Himalayan coat pattern in rabbits, 116, 124, 258–260
Histones, 211
Holandric genes, 107–108
Homozygosity, 45
Honeybee, sex determination in, 91
Hormones, and sex determination, 91–94
Hutt, F. B., 144
Hybrids, interspecies, 38
Hydrogen bomb, 247
Hypertrichosis, 108

Icthyosis, in man, 108, 225
Ingram, V. M., 195
Intelligence, in twins, 267
Inter se cross, 41
Interference, in crossing over, 147–148
Intermediate genes, 50
Interphase, 25–26, 28, 29
Intersex, in *Drosophila*, 172
Inversion, chromosome, 158–159, 167
Isotopes, radioactive, 228, 236, 238

Japan, atom bomb studies of, 253–254
Jaundice, hemolytic, 83
Jimson weed, 161–162
Johansen, Wilhelm, 22
Jorgensen, Christine, 93

Kappa bodies, 190–191
Keratoma dissipatum, 108

Kerry cattle, 54, 57
Kinetoplasts, 189
Klinefelter's syndrome, 177–178

Lamarck, Jean, 17–18
Larkspur, 217
Leeuwenhoek, Anton van, 14–15
Lethal genes, 52–55, 215, 220–223
Lewis, E. B., 152
Linked genes, 141–152
Liverworts, sex determination in, 175
Lymantria, 174
Lysenko, Trofim, 18–19

Malaria, and sickle-cell trait, 197
Manx cats, 51
Matrix, 26–27, 29
Maupertuis, Pierre, 16–17
Meiosis, 31–40
 in animals, 32–35
 gene assortment and, 37–38
 in interspecies hybrids, 38–40
 in plants, 39–40
Melandrium, 176
Mendel, Gregor, 21–22, 41–45, 59–61
Messenger RNA, 207–208
Metaphase, 26, 27, 28, 29
Mice, use in genetic studies, 5
Mitochondria, 189
Mitosis, 25–31
 anaphase, 26, 27–28
 interphase, 25–26, 28, 29
 metaphase, 26, 27, 28, 29
 prophase, 26–27, 28, 29
 telophase, 26, 28–29
MN blood antigens, 133
Mongolism, 178
Monohybrid cross, 41–55
 diagramming, 48
 intermediate expression of genes in, 50–52
 mechanisms of, 45–47
 Mendel's experiments with, 41–45
 test cross, 48–50
Monosomic plants, 161
Morgan, T. H., 22, 100, 213
Mulatto, 120

Muller, H. J., 223, 227
Multiple exostoses, 263
Multiple genes, 114, 116–121
Mutations, 20, 210, 213–232
  artificial induction of, 227–229
  detection of, 218–223
  DeVries' theory of, 20–21, 213
  doubling dose in, 247–249
  frequency of, 223–227
  harmful, 215–216
  kinds of, 215
  natural, 229–231
  reverse, 215
  somatic, 216

Natural selection, 19
Neel, James V., 102, 226, 254
Neurospora, cytoplasmic inheritance in, 191
  enzymes in, 200–203
  mutation detection in, 222, 223
Neutron, 236
Newman, H. H., 266
Nilsson-Ehle, H., 22
Non-disjunction, in *Drosophila*, 160, 170, 172

Octaploid cells, 163
Onion mitosis, 26
Oögenesis, 35–36
Operator genes, 211
Operon, 211
Otosclerosis, 44

Pangenesis, 19–20
Paramecium, 190
Particulate radiation, 236
Pascal's triangle, 76
Paternity, determination by blood type, 131–132
Pauling, Linus, 195
Peas, Mendel's experiments with, 21–22, 41–45, 59–61
Pedigrees, in genetics, 7–8
Pelger anomaly, 225
Penetrance, 262–263
Pes cavus, 52

Phenocopies, 263–265
Phenotype, 45–46
Phenylalanine metabolism in man, 203–206
Phenylpyruvic idiots, 204
Plastids in plants, 189–190
Pneumococcus, 180
Polygenes, 114, 116–121
Polyploid cells, 162–163
Polyribosomes, 207
Polysomic plants, 161
Position effect, in pseudoalleles, 150–152
Preformation theory, 15
Primroses, evening, DeVries' experiments with, 20–21, 213
Probability, 70–82
  analysis of, 77–82
  binomial method in, 74–76
  coincident happenings and, 71–74
  genetics and, 70–71
Prophase, 26–27, 28, 29
Pseudoalleles, 137, 150–152, 188
Pseudohermaphrodites, 94–95
Puffing phenomenon, 211
Punnett, R. C., 22, 60, 141, 142
Punnett's squares; *see* Checkerboard method
Pure-line concept, 22
Pythagoras, 13

Quantitative characteristics, and multiple genes, 118–121

Radiation, 233–256
  atomic, 246–247, 253–258
  biological effects of, 241–247
  genetic effects of, 247–251
  hazards, reduction of, 233, 254–256
  measurement of, 239–241, 252–253
  medical use of, 243–244
  protection from, chemical, 250
  sources of, 233–238
  thirty-year gonadal dose, 253
  tolerance doses, 251–253
Radioactive isotopes, 228, 236, 238

Ratio theory of sex determination, 171
Recessiveness, 42–43
Recon, 188
Regulator genes, 211
Retinitis pigmentosa, 107
Retinoblastoma, 225
Ribosomes, 207–208
Rh blood antigens, 133–138
Rickets, 269–270
RNA, 207–212
Roentgen, 233
Russell, William I., 248

Salivary gland chromosomes, in *Drosophila*, 163–169
Schultz, J., 172
Secretor trait, 132–133
Segregation, Mendel's law of independent, 59–60
Sex chromosomes, 87, 94, 170–178
Sex determination, 86–98
　by chromosomes, 87–91, 170–178
　hormones and, 91–94
　ratio in, 96–98, 171–172
　separation of sexes in, 86–87
　sexual bipotentiality in, 86
Sex-influenced genes, 110–111
Sex-intergrades, 94–96
Sex-limited genes, 108–110
Sex-linked genes, 100, 108
　in *Drosophila*, 148–150
Sex, predetermination of, 98
Sex ratio, 96–98, 171–172
Sex reversal, 91–93
Sheep, ancon, 213, 217
Shull, William J., 226, 254
Sickle-cell anemia, 56, 193–198, 231
Skin color, in man, 120
Somatic mutations, 216–217
Sperm, human, 32–37
Spermatogenesis, 32–35
Spindle fibers, 27
Stadler, L. J., 226, 227
Standard deviation, 122–123
Standard error, 77–79
Stature; *see* Height
Stern, Curt, 120
Stiles, Karl A., 261
Strontium 90, 247
Structural genes, 211
Sutton, W. S., 141
Swammerdam, Jan, 15
Sweet peas, *see* Peas
Synapsis, 32
　somatic, 164

Teeth, defective enamel of, 106
Telophase, 28–29
Test cross, 48–50
Tetraploid cells, 162
Tongue rolling, 43, 230
Transduction, in bacteria, 182–183
Transfer RNA, 207–208
Transformation, in bacteria, 182
　in ducks, 184
Translocation, chromosome, 159–160, 167
Triploid cells, 162
Trisomic plants, 161
Tryptophan synthesis in *Neurospora*, 201–202
Turner's syndrome, 178
Twins, environmental studies of, 265–270
　origin of, 29
Tyrosinosis in man, 204

Ultraviolet rays and mutations, 228–229
Uranium, 234
Use and disuse theory, 17–18

Virus, and gene structure, 180–182
Vitamin requirements, 270

Warmke, H. E., 176
Watson Crick theory, 185–187
Weismann, August, 19–20
Westergaard, M., 176
Wiener, A. S., 136
Wilson, E. B., 22
Wolff, Kaspar, 17

Xanthomatosis, 52
X rays, 227–228, 233–234

Yeasts, 191

Zygote, 23